Our Superstring Universe

Our Superstring Universe

◆

Strings, Branes, Extra Dimensions and Superstring-M Theory

L.E. Lewis, Jr.

iUniverse, Inc.
New York Lincoln Shanghai

Our Superstring Universe
Strings, Branes, Extra Dimensions and Superstring-M Theory

iUniverse, Inc.

For information address:
iUniverse, Inc.
2021 Pine Lake Road, Suite 100
Lincoln, NE 68512
www.iuniverse.com

ISBN: 0-595-27589-3

Printed in the United States of America

Contents

Acknowledgments

Cover Photographs

Front Cover: The Eskimo Galaxy (NGC2392). NASA, ESA, Andrew Fruchter and the ERO Team (STScI)

Text Illustrations

Figure 1.13: NASA, the NICOMOS Group (STScI/ESP) and the NICOMOS Science Team (University of Arizona)

Figure 1.15: Hubble Heritage Team (AURA/STScI/NASA)

Figure 1.16: STScI, NASA, Brad Whitmore

Figure 1.17: STScI/NASA

Figure 1.19: STScI/NASA

Figure 3.1: STScI/NASA

Figure 3.5: NASA/CXC/SAO/A.Vikhlinin

Computer Generated Graphics

The text illustrations, other than those credited above, were computer generated by the author with the encouragement and assistance of Computer Specialist, David L. Lewis.

Concept Development Discussions

My special thanks to Curtis Hohman and David L. Lewis for participating in the many discussions that were so very helpful in selecting and putting together the concepts most basic to superstring-M theory. Thanks, too, for proofing the manuscript.

Understandability Evaluation

Thanks also to the following for helping me, through discussions, to organize and "level" the writing for the intended reader: Todd V. Anderson, Brian Leather-wood, Brian L Masters, Dana D. Masters, Harry B. O'Daniel, Scott Sartain, and Christopher Tucker.

My sincere appreciation to all of the above,

L. E. Lewis, Jr.

June 2003

Preface

Our Superstring Universe was written for readers who have a keen interest in understanding our universe and the stuff of which it is made. It is hoped that this book will provide a starting point and foundation for readers new to the subject. For those with some knowledge of particle physics, perhaps it will provide an overview of string theories and serve to stimulate awareness of their potential.

Our Superstring Universe is intended also as an introduction to what may soon be the most important theory of science...*ever*. It will be important to everyone, not just scientists. It is the first unified physics theory of everything in our universe...the superstring—M theory. We are tentatively referring to this theory as the superstring-M theory because, at this time, string and M theories are in a process of modification, refinement and combination.

The more recent M theory has stimulated activity in string theory because it broadens string theory and shows promise of clearing up some problems. The reader should realize that some of the information about superstring-M theory (SS-M theory) will become out of date, rather quickly. However, suggestions for keeping updated are given in the back of the book in the form of websites and science magazines. In addition, for those of you who wish to continue the study of SS-M theory a few selected next-level books related to the subject are listed on page 160.

The organization of Our Superstring Universe is primarily one of logical, sequential concept development. Concepts, definitions and information introduced early are used to build the more difficult concepts throughout the book. Some information is also in chronological order. This is obviously necessary in the evolution material. However, the chronology is interrupted early, for a few pages, in order to develop some basic concepts and definitions essential to understanding of the evolution sequence.

The primary objective of *Our Superstring Universe* is to introduce and build a basic understanding of superstring-M theory. The origin and evolution of matter coincide with the origin and evolution of our universe. This is obvious, of course, because there would be no universe (as we know it) without matter; and it is one reason the concepts of particle physics are interwoven with those of cosmology.

In addition, cosmology provides excellent examples of the relationships between forces, matter particles and the events and theories of our universe. Moreover, astrophysics and cosmology are very rapidly becoming interrelated with particle physics. As you will note, much of the current progress in the refinement of the theories of matter is coming from cosmological discoveries.

A couple of unique concepts are used in the book because they are very helpful in explaining string theory. The first is the introduction of the *space particle*. A space particle is defined as the same tiny curled-up shape that is fundamental to string theory. In our discussions, the curled-up shape is also considered to be the *elementary component of space…the space particle.*

Most of the literature states that the size of the shape is about the Planck length. The literature also states that they exist *at every point in space,* even though they have a tiny size. Since they have size, they have volume and must occupy space. All of this creates a bothersome problem: How can shapes be at every point in space if points have zero size and shapes have a size greater than zero?

If curled-up shapes are viewed as space particles that have a tiny volume, they can't be at every point in space. However, they can *fill the volume of the space of our universe.* This would be somewhat like the molecules of air fill the volume of a balloon. Since our universe began expanding at the time of the big bang and is still expanding, space particles must have been created continuously since the big bang. In this way, the volume of our universe is kept full of space particles. Again, this is somewhat analogous to the way blowing molecules of air into a balloon keeps the balloon full of particles (molecules) as it expands.

This continuous creation of space is consistent with current theories, which assume that *space did not exist* (except as encoded in the nugget) until it began forming at the birth of our universe. Moreover, it is also generally accepted that the constant expansion of our universe *requires* the continual creation of space. If space particles are assumed, as defined, we have a particle that can be thought of as the *quantum* of space. Is this "particle of space hypothesis" likely to be the way space really is formed? Possibly, but probably not because not enough is known about the properties of space at this time. In any case, it is hoped that the concept of space particles will aid the reader in developing an understanding of our universe and superstring-M theory.

The second unique concept arises from an interpretation of a feature of SS-M theory. That feature is the *holographic principle,* which seems to imply and allow a concept generally considered to be outside the realm of science. It is the concept of a human afterlife. A hypothetical mechanism for creating an afterlife based on certain aspects of SS-M theory is explored briefly. This is primarily intended as an example of the expected breadth and power of SS-M theory to explain many of the mysteries of nature.

Some of the illustrations in the book are intended to aid understanding of a specific concept or definition. Others restate the concept to reinforce understanding. In addition, a few provide additional information on a topic. To offset the lack of color photographs (too expensive) you can find many thousands of color photographs on NASA or NASA-related websites.

Before beginning Section 1, it is suggested that the reader examine the *Learning Tips, Websites* and *Glossary* in the back of the book.

Section 1

The Origin and Evolution of Our Universe

1

The Origin of Our Universe

THE BIG BANG

Time Interval: zero to 10 to the -43rd power of a second
Temperature: 10 to the 32nd power degrees K

A little more than 14 billion years ago our universe existed only as an unbelievably tiny ultramicroscopic nugget .00000000000000000000000000000001 of a centimeter in diameter. Our universe-to-be was encoded within the nugget. The nugget was all that existed. Energy, matter, space and time as we know them, did not exist. The nugget was suspended in a "void of nothingness". Inside the nugget were ten extremely compacted curled-up spatial dimensions and one time dimension. The temperature and density were colossal.

Suddenly…at time zero…the big bang! Time began ticking as a cataclysmic explosion unleashed hot, chaotic "plasma" into the void of nothingness. This frenetic, hot, plasma-like, stringy space interacted violently. Three of the compacted spatial dimensions worked free of their bonds and inflated rapidly. This was the beginning of the three-dimensional, visible portion of our universe. All of this happened in an infinitesimal fraction of a second.

Figure 1.1 The Nugget Explodes. Suddenly, the infinitesimal nugget burst, and our universe was born. That was about 14 billion years ago, and the energy of the blast is still driving the continued expansion of our universe. It seems that somehow that tiny nugget was encoded with all of the information necessary to create our universe through a process of continuing evolution.

Space Particles

Although they did undergo violent interaction, the other six or seven dimensions did not break their wrapped strings and remained curled-up. However, they did go through thousands of frenetic, space-tearing changes from one shape to another. Then, one special design of six or seven dimensions evolved through a natural selection process and became stable. As we will see later, these space particles (or shapes) are the objects in which the strings of string theory reside. See Figure 1.2. The space particles are so tiny that if one were magnified to the size of a school bus, at the same magnification, an atom would be about the size of an average galaxy.

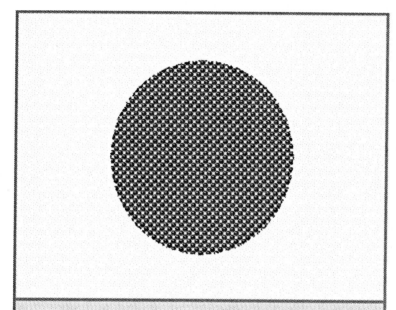

Figure 1.2 Space Particles. Space particles are pictured as they might look magnified trillions and trillions of times. This is, of course, only a simulation of a cross section of space showing how the space particles fill the volume of 3-dimensional space.

In our discussions, these extremely tiny shapes are also considered to be the fundamental *particles of space*. The *space particles* replicate from the energy of the big bang and are carried along within the constantly expanding three-dimensional visible portion of space. As our universe continues to expand, space particles are created to fill-in its increasing volume. (See the Preface for a more detailed discussion of space particles.)

SEARCHING FOR FUNDAMENTAL UNITS

Space and Energy

It is now about 10 to the negative forty-third power of a second since the big bang. Let's pause, to see if we can discern clues to the nature of the "stuff" that exploded from the nugget. What were the first fundamental components of our universe? First, note that the nugget was described as containing *space*...extremely dense, hot space of ten curled-up dimensions, wound-up like a tightly wound spring.

The extreme density, (compaction) is indicative of enormous *potential* (stored) *energy in the nugget*. The extremely high temperature, about ten trillion trillion times the temperature of the core of the sun, also indicates the enormous amount of energy present at the big bang. *Space* and *energy* must have been two of the fundamental components of our baby universe.

Motion, Change and Time

Remember that in this theory it is assumed that *nothing* existed outside the nugget. At the moment before the big bang there was *only* the nugget suspended in a void of nothingness. When the nugget exploded, the compacted spatial contents were hurled violently in all directions. The nugget and the void around it changed drastically as the stuff of our universe *moved* outward.

Since change occurs as a result of motion through space, we intuitively realize that *change* must occur during an interval of *time*. Although there is much we don't understand about time, *time* must be a crucial component of our universe. In addition, it is generally assumed that time began with the big bang.

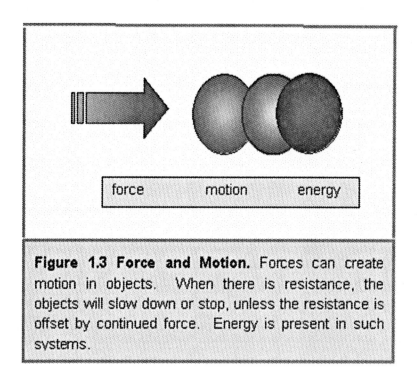

force   motion   energy

Figure 1.3 Force and Motion. Forces can create motion in objects. When there is resistance, the objects will slow down or stop, unless the resistance is offset by continued force. Energy is present in such systems.

To this point, we have identified space, energy, change, motion and time as fundamental components of our universe. Notice, however, that change and motion are not as fundamental as the other three. Since change in the state of motion (such as a push on an object, causing the object to move as shown in Figure 1.3) defines energy, change and motion fall under the definition of *energy*. That makes three fundamental components: *space, time* and *energy*.

Matter

We have been able to describe the earliest infinitesimal moment of our universe without evoking matter. This is not surprising since matter didn't exist, as we know it, during that moment of time. The density and temperature were many orders of magnitude greater than even that of the plasma state of matter in the sun. We did, however, describe the stuff exploding out of the nugget as extremely hot, "plasma-like, stringy space".

There must be a close relationship between *space* and *matter*. The "stuff" must be a precursor of matter. Recall that space, as we have defined it, consists of minute particles of six or seven dimensions, filling an ever-expanding volume of three dimensions.

Space-Time

Time is difficult to conceptualize, partly because we can't see or feel time. However, we do have "a sense of time". We also use operational definitions of time based on "the uniform motion of some object through space". For example, our year is based on one revolution of the earth around the sun; and our day is based on one rotation of the earth.

As we will see soon, in string theories the vibrating object is a one-dimensional string that can be thought of as a "sliver of space". Using this concept, we can see that time can be described as the uniform motion (vibration) through space of a string (made of space). This description of time gives us a rough idea of the close relationship between time and space.

All of the above may give the impression that space is more fundamental than time. In any case, we can see that space and time are closely interrelated. Space, however, does seem to be a notch above time; and energy seems to be a notch above matter even though one couldn't exist without the other.

Early in the last century, Albert Einstein demonstrated the close relationship between space and time through his mathematical theories of relativity. Physicists have found relativity theories to be central to the current concepts of our universe. Even without the mathematics, we can see why space and time are conceptually interrelated and can be denoted as a *dual* component of our universe as: *space-time*.

Matter-Energy

In string theories, matter is produced by vibrating strings; and vibration is a form of energy. This indicates that there is an interrelationship between matter and energy. In addition, Einstein demonstrated the relationship between matter and energy in his familiar equation: $E=mc^2$ [Energy=Mass (of matter) X the speed of light *squared*]. Nuclear energy, which is produced by turning matter into energy, is proof of the equation. Thus, we have very good reasons to denote matter and energy as matter-energy. That leaves us with two *dual* fundamental components of our universe: *space-time* and *matter-energy*.

SYMMETRY

In addition to the above prime components, there is a recurring structural feature that appears in our universe, from the core of the nugget to the nuclei of atoms and the vast expanses of our universe. It is *symmetry*. In particle physics symmetry is a spacetime property of physical systems that causes them to *look the same* when the system is *transformed* in a specific way.

A butterfly, the human face, the human body and a leaf are examples of a familiar kind of symmetry. They exhibit *bilateral* or *mirror symmetry*. The mirror images of these objects look the same (although not perfect in these cases) when transformed by a mirror placed along a line through their centers.

Another kind of symmetry is called *rotational symmetry*. A sphere is *rotationally symmetrical* because it looks the same when rotated on any axis, or when viewed from any angle. See Figure 1.4. Other examples of rotational symmetry are a baseball, a globe and a spinning top.

Symmetry can be represented mathematically, and it appears in the formulas of superstring theories. We will avoid the mathematics by using examples and analogies in our discussions, and we will only refer to the mathematical formulas now and then. Symmetries are present in all kinds of matter, from the smallest particles to the largest galaxies. Superstring theories indicate that the nugget had the highest degree of symmetry of anything...*ever*.

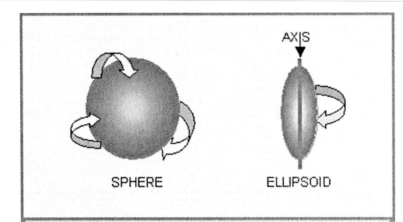

Figure 1.4 Rotational Symmetry. An object has rotational symmetry if it looks the *same* when rotated on *any* axis or when viewed from *any* angle. The sphere always has rotational symmetry. The ellipsoid does not. However, the ellipsoid does have rotational symmetry about the axis shown. Matter particles, such as the electron also rotate (spin) around an axis.

Symmetry Breaking

The role of symmetry in the evolution of our universe is crucial. Often it involves the process of *symmetry breaking*. A good example is the *change of phase* that occurs when liquid water is cooled to a critical temperature. At a molecular scale liquid water appears essentially the same when viewed from any angle as it is heated or cooled.

The speed of the random motion of the molecules of water does increase and decrease when heated or cooled; but the general arrangement within the volume remains the same. Therefore, liquid water has rotational symmetry. However, when it is cooled to zero degrees Celsius, liquid water makes a phase transition and *changes to the solid phase*: ice or snow crystals. When water changes from the liquid phase to the solid phase it no longer looks the same when its structure is viewed from *any* angle.

This is an example of *symmetry breaking* since the water no longer has the rotational symmetry it had in the liquid phase. This *does not* mean that water in the solid phase has no symmetry.

Figure 1.5 Symmetry Breaking and Building. Water has rotational symmetry. When water changes phase from liquid to solid its rotational symmetry is broken. This is *symmetry breaking*. This is true even though it still has other kinds of symmetry when in the form of ice. When ice or snow melt *symmetry building* occurs since the water regains its original symmetry. In the scene above ice and snow are changing phase to water. This is an example of *symmetry building*.

Although they don't have the same rotational symmetry, the arrangements of particles in an ice crystal or in a snowflake do have other kinds symmetry. For example, the snowflake has rotational symmetry about its center point in a plane. However, it doesn't have the same kind of rotational symmetry that water has.

Symmetry Building

Symmetry building is another process important in the evolution of our universe. It can be illustrated by the transformation of ice into liquid water by heating. Obviously, this is the reverse of symmetry breaking. See Figure 1.5. The change of phase gives the particles (molecules) of water rotational symmetry. If we examine the evolution of our universe starting at the present, instead of at the big bang, we find symmetry building at each step where we find symmetry breaking when we start at time zero.

There are times when symmetry building also occurs during our "zero to the present" discussion. One example is the change of phase that occurs, as the particles in a blob of gas are pulled closer and closer together by gravity. This compaction can eventually raise the temperature, until the core of the blob reaches the plasma phase. This extremely dense, hot *plasma phase* initiates *nuclear fusion*, which is the source of the high temperature of stars and their radiation. You will note many examples of both symmetry building and symmetry breaking in our continued discussion of the evolution of our universe.

Supersymmetry

An awareness of the central role of symmetry in our universe led to the realization that symmetry is an important recurring theme in nature. This realization of the central role of symmetry in nature led to the search for other forms of symmetry. As a result, *supersymmetry*, a form of symmetry involving "quantum spin", was added to both the standard theories and the superstring theories. To reflect this, their names were changed to the supersymmetric standard theories and the superstring theories, respectively. We will learn more about supersymmetry and its role in particle theories later.

Superstring Theories

The initial string theories of matter were improved as various numbers of dimensions and new kinds of symmetries were tried. Several surprises emerged through the years. One especially important surprise was the unexpectedly close relationship between matter particles and force particles. Both emerge from superstring theories; and both are produced by vibrating strings.

Matter particles have mass; messenger (force) particles do not except for the weak force particles. There are a half-dozen viable superstring theories and a theory called M theory. No one seems to be sure what the M originally stood for in M theory. Perhaps it was "membrane", since it allows for "branes", as we shall see later.

2

The Evolution Continues

BACK TO THE BIG BANG

Time Interval: zero to 10^{-43} of a second ATB
Temperature: 10^{32} degrees K (See the *Glossary* for an explanation of "powers of ten".)

Finally, we are ready to resume discussion of the cosmology of our universe. Remember, the big bang had just occurred; and a colossal explosion unleashed an unbelievably hot, roiling, rapidly expanding "stuff"...the precursor of space, time, matter and energy. This first symmetry breaking occurred, as three of the compacted spatial dimensions broke free of the strings wrapping them. (We will see, in Section 2, that strings can wind around dimensions.) These dimensions became the three-dimensional portion of our visible universe.

The other six or seven spatial dimensions remained compacted and became the ultramicroscopic curled-up portion of our universe. Although they struggled to break their bonds, the curled-up dimensions remained compacted and evolved into a special shape. Particles with this special shape became the particles of space of our discussions.

GLUONS ARE RELEASED

Time Interval: 10^{-42} to 10^{-37} of a second
Temperature: 10^{28} degrees K

Our universe continued to expand and cool causing symmetry breaking to condense the *strong force* (gluons) out of the unified force plasma. The strong force will soon be needed to bind quarks together to form protons and neutrons.

A COLOSSAL EXPANSION

Time Interval: 10^{-37} to 10^{-34} of a second
Temperature: average about 10^{27} degrees K

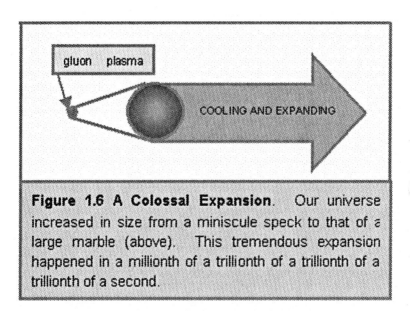

Figure 1.6 A Colossal Expansion. Our universe increased in size from a miniscule speck to that of a large marble (above). This tremendous expansion happened in a millionth of a trillionth of a trillionth of a trillionth of a second.

Suddenly our universe shifted gears and stepped on the gas. During an extremely tiny flicker of time, our universe expanded by 25 orders of magnitude! The infinitesimal "seed" increased in size from that of an invisible ultramicroscopic particle to the size of a large marble. See Figure 1.6. (The sphere is the actual size of our universe after the expansion.) That may not seem like such an enormous expansion, but it is equivalent to a pea expanding to the size of a typical galaxy.

That was a dramatic moment in our baby universe. Before the enormous inflation the matter-energy plasma was extremely dense. The closeness of all parts of the plasma and the extremely rapid inflation assured that all parts of our universe would always have a common temperature. This is similar to the way objects in a room reach, with time, the same temperature. The miniscule "clumps" of hot plasma retained a common temperature while moving apart during the extremely rapid expansion.

The standard cosmological theory begins with a *zero* size (instead of a tiny nugget) and *infinite* density and temperature. In addition, it has uniform expansion from the time of the big bang, and it does not have the brief colossal expansion. Major modifications were made to the standard theory by the application of relativity and quantum theories. Most cosmologists accept the modifications. The modified theory is called the *inflationary cosmological theory* and is the one we are following.

Physicist Alan Guth proposed the brief colossal expansion, in 1979, as the *grand inflation* theory. Since that time, a number of interesting modifications of the theory have been made. Modifications continue, and you should hear about them from time to time. Among the most interesting new ideas are those that would replace the nugget with a "random energy fluctuation in the void of nothingness". We will examine this idea further in Section 3.

THE WEAK FORCE AND QUARKS APPEAR

Time Interval: 10^{-34} to 10^{-3} of a second
Temperature: Average about 10^{15} degrees K

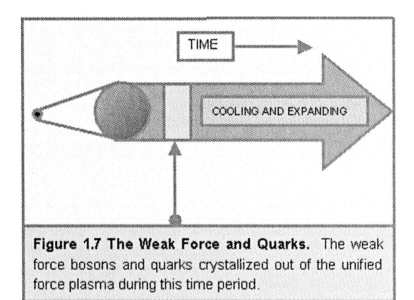

Figure 1.7 The Weak Force and Quarks. The weak force bosons and quarks crystallized out of the unified force plasma during this time period.

During the last part of this time interval, expansion, cooling and symmetry breaking began condensing (or crystallizing) the *electroweak force* particles (*bosons*) out of the unified field force. (Later, bosons will be needed in the nuclear interactions between certain particles.)

Next, the first matter particles, *quarks*, were released from the plasma matter-energy pool. As quarks became available, they formed groups of three, bound tightly together by gluons. See Figure 1.8. (Recall that gluons are the strong force particles, isolated earlier from the unified force plasma.) The three-quark combinations will become *neutrons* and *protons*, which can combine in various combinations to form the nuclei of the atoms.

PRIMORDIAL NUCLEOSYNTHESIS

Time Interval: About 3.5 minutes
Temperature: Average about 10^9 degrees K

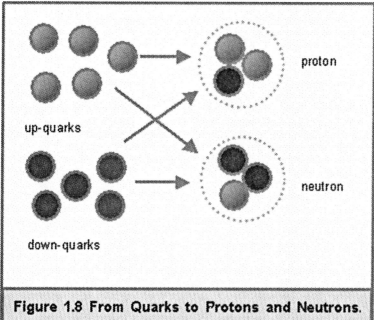

Figure 1.8 From Quarks to Protons and Neutrons. Two up-quarks, one down-quark and gluons make a proton. Two down-quarks, one up-quark and gluons make a neutron. Protons carry one positive charge. Neutrons have no charge (zero).

Finally, a time interval we can understand. It is still hard to believe how so much could have happened during the tiny fraction of a second from the big bang until now. Perhaps, it has something to do with the nature of spacetime. After all, those events were pretty weird. In any case, the protons and neutrons have increased in number, forming the nuclei (positively charged ions) of future atoms.

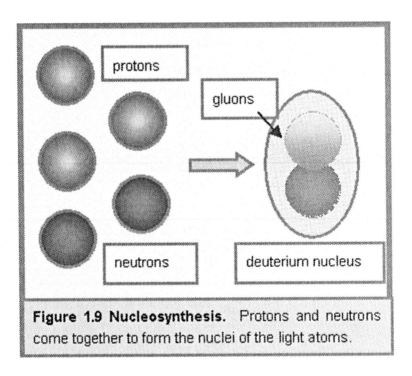

Figure 1.9 Nucleosynthesis. Protons and neutrons come together to form the nuclei of the light atoms.

Our universe continued to expand, the temperature continued to cool; and *protons* and *neutrons* formed stable nuclei, held together by strong force gluons. See Figure 1.8. At this point, there were no electrons (which are negatively charged). The only matter particles present were the positively charged nuclei. Their positive charges come from the proton(s) in the nucleus, which carry one positive charge each. Such charged particles are also called *ions*.

The nuclei formed were those of the lowest mass (weight) primarily hydrogen, some helium, a bit of deuterium and lithium. (Deuterium is often referred to as heavy hydrogen. It has a nucleus of one proton and one neutron.) At this time, our universe was a dark, seething swarm of energetic positive nuclei awaiting electrons to make them into atoms.

THE DARK ZONE

Time Interval: About 300,000 years
Temperature: 10^8 to 10^5 degrees K

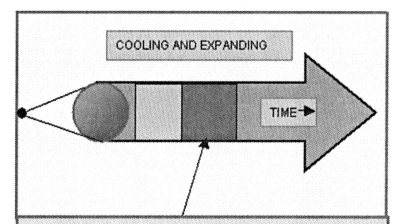

Figure 1.10 The Dark Zone. During this era of about 300,000 years, great numbers of energetic electrons were formed. Gradually, the electrons slowed enough so that the positively charged nuclei could capture them. This cleared our universe of most of the charged particles (ions) and freed light to travel through space.

THE FIRST ATOMS

As expansion and cooling continued, *electrons* and other highly energetic particles such as *neutrinos* were formed. Since the matter-energy plasma of this era was dense with positively charged nuclei, negatively charged electrons and other energetic ions, our universe was dark. This is because ions do not transmit light well.

After the temperature dropped a few thousand degrees, the electrons and positive nuclei slowed enough to allow them to merge, forming the first *atoms*. See Figure 1.11. As the positive nuclei and negative electrons combined to form atoms, the ions were neutralized; and our universe became transparent to light.

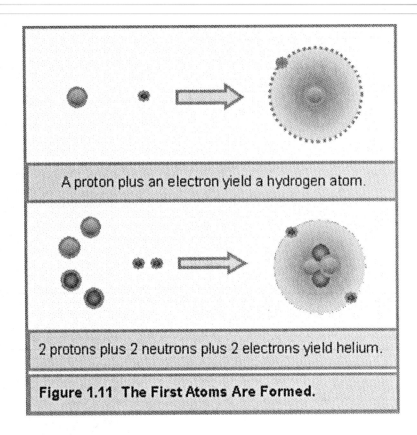

A proton plus an electron yield a hydrogen atom.

2 protons plus 2 neutrons plus 2 electrons yield helium.

Figure 1.11 The First Atoms Are Formed.

When the first atoms were formed, and as they continued to form for millions of years, they were still very energetic. Their high temperature kept them apart until they cooled enough to begin to clump together. Slowly, the atoms were drawn closer together; and the stage was set for some big time symmetry *building*...the birth of stars.

FROM ATOMS TO STARS

Time Interval: About 200 million years ATB to the present
Temperature: Average about 100,000 degrees K

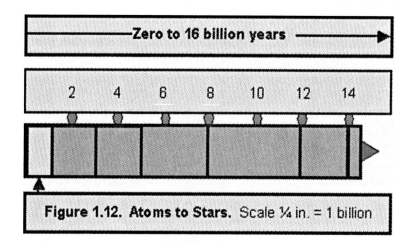

Figure 1.12. Atoms to Stars. Scale ¼ in. = 1 billion

As the frenetic dance of the newly formed atoms cooled, their motion slowed; and the gravitational force between atoms gradually pulled them together. Slowly, molecules and clusters of molecules formed. As the clusters grew larger, their mass and gravitational force increased. This caused the larger clusters to pull smaller ones toward them and merge.

This process continued for millions of years as gigantic blobs of gas, (mostly hydrogen) formed and populated this era of space and time. As a blob became more and more massive, its gravitational pull became colossal. This forced the atoms and molecules of the blob closer and closer together, eventually crushing the gaseous sphere to great density and temperature. As the temperature rose, symmetry building really set in. The gases became more and more frenetic, molecules broke down, electrons were stripped from atoms, intense radiation was emitted; and eventually when the matter in the core reached the critical density and temperature, *nuclear fusion* ignited. In this way, the giant blobs of gas were transformed into stars. See Figured 1.13. Their fusion furnaces, transmuting hydrogen into helium, would keep some of them burning for billions of years. However, a few (supernovae) will have shorter lives, which end in brilliant explosions.

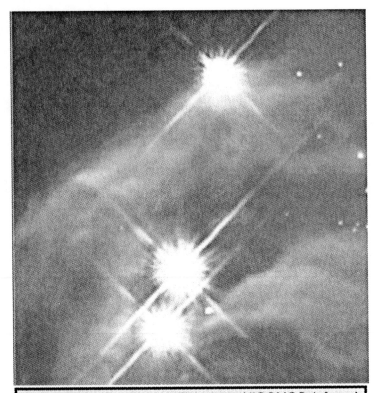

Figure 1.13 Star Birth. This is a NICOMOS infrared image of a dense portion of the tip of the Cone Nebula. Infrared waves penetrate the layers of dust well enough to indicate that the tip is a turbulent star-forming region. It is about ½ light year across; and the whole nebula is about 7 light years across. Supernovae explosions appear to have contributed to star formation by providing huge blobs of dense hot gases. Gravity draws the molecules of the gases closer and closer together, until the core reaches the critical density and temperature needed to ignite fusion, and form a star.

The first stars suddenly flashed into existence about 200 million years ATB. This was the beginning of the many trillions and trillions of stars to come. Soon supernovae, quasars, black holes, galaxies and galactic clusters began forming. The creation of our universe, as we know it, was really cooking.

From Stars to Galaxies

Time Interval: From 200 million years ATB to the present
Temperature: In the beginning several thousand degrees K

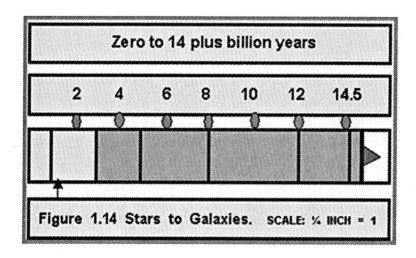

Zero to 14 plus billion years

2   4   6   8   10   12   14.5

Figure 1.14 Stars to Galaxies. SCALE: ¼ INCH = 1

Galaxies

As trillions of stars were born during the previous epoch, the enormous gravity of the stars began drawing them together into *galaxies*. Over millions of years, the gravity gathering process continued, as galaxies of varying sizes and shapes were created. Recent observations indicate that the earliest galaxies were globular clusters of only a few hundred thousand stars. Furthermore, it now appears that these galaxies originated from the debris of massive supernovae explosions.

This means that many of the first stars were *supernovae* with masses as much as 300 times that of our sun. Although the early galaxies were relatively far apart, the gravity of a galaxy of billions of stars was huge. It was enough to cause some galaxies to merge and form larger galaxies. Close encounters also changed the shape and motion of some galaxies. There are elliptical, spherical and spiral galaxies, and variations of these shapes such as spiral galaxies with thin spiral arms.

The Hubble space telescope (HST) shows evidence of strong interaction between about one-third of the early galaxies. It appears that many galaxies merged to form larger galaxies. In addition, the HST shows that groups of galaxies formed into *galactic clusters*. It now appears that supernovae, galaxies and galactic clusters formed much earlier in the evolution of our universe than previously thought…beginning soon after the first stars were born.

Many of the descriptions of the events along the preceding timelines of our universe were derived from theoretical conclusions. Now, however, as illustrated above, astrophysics has entered a period of high technology. The HST (see Figure 1.19) and a number of other specialized observatories can be described as photographic time machines. After all, the HST is actually taking pictures of objects, *as they were* up to 14 billion years ago, near the beginning of our universe.

Some photographs show baby galaxies in early stages of formation, pretty much as expected. However, very recent photographs show enormous *quasars;* and it now appears that these very brilliant objects also were formed very early in the evolutionary sequence of our universe.

Through some 14 billion years to the present, galactic clusters, their stars, interstellar gases and dust continued to change. The ever-present force of gravity, both within galaxies and between them, is still a major factor in causing change. As indicated above, recent deep space observations appear to indicate that somehow galactic clusters formed amazingly soon after the big bang. This would have been too early for the gravity of ordinary matter alone to cause cluster formation. As we shall see later, *dark matter* is believed to have played a crucial role in organizing the early galaxies into clusters and holding them together.

Figure 1.15 A Spiral Galaxy. This is an HST photograph of spiral galaxy NGC 4414. It is somewhat typical of this type of galaxy. Its center region contains primarily older red and yellow stars. In the outer areas of the spiral arms are bright blue stars. Some are bright enough to be seen individually. Since the spiral arms are rich is dust and gases, we can expect stars to continue to form there for some time. There are a number of variations of the spiral shapes of galaxies.

Maturing Galaxies

Time Interval: About 300 million years to the present
Temperature: At present, the overall average temperature of our universe is a little more than 2 degrees K

Another kind of change was occurring inside stars, during this period. As a star burns hydrogen, the density and temperature eventually decrease to a level that can no longer sustain fusion, and the star dies. In some cases the star rejuvenates to shine again. New stars also form as the gases and dust of a galaxy are compressed enough to ignite the fusion reaction. Sometimes two or more galaxies merge. This process can happen more than once.

Colliding Galaxies

These collisions (merging) can occur on a grand scale. See Figure 1.16. For example, our Milky Way Galaxy is currently in the process of swallowing Sagittarius, a nearby dwarf galaxy. In only 200 million years from now, the ten million stars of Sagittarius will have been added to the 200 billion stars of our Milky Way galaxy.

At the same time, another even greater drama is taking place in our region of the universe. Our Milky Way Galaxy and the Andromeda Galaxy are approaching each other at 300,000 miles per hour. Don't worry, they are still about 13 billion billion miles apart, so they won't merge for about 5 billion years.

When they do merge, millions of new stars will ignite as huge blobs of gas and dust are compressed and heated by the collision. Notice that some of the enormous amount of interstellar gas and dust in our universe is transformed into stars in such events as these.

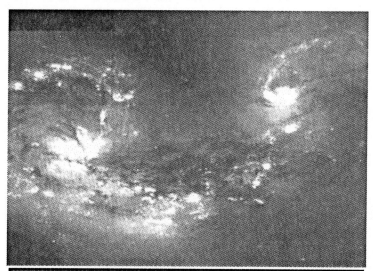

Figure 1.16 Colliding Galaxies. This HST photograph shows the collision of the Antenna Galaxies. The cores of the twin galaxies are the two bright areas to the right and left of center. A band of chaotic dust stretches between the cores. As blobs of these gases combine to form larger and larger masses, they reach the critical density and temperature necessary to ignite fusion; and new stars are born. Already, over a thousand *clusters* of bright, young, hot stars have formed. These clusters highlight the spiral arms of the twin Antenna Galaxies.

Seeding the Galaxies

As indicated earlier, when the matter of our universe was initially created it consisted primarily of hydrogen. Helium was present at about 13 percent, and there were small amounts of deuterium (heavy hydrogen) and a bit of lithium. Medium weight elements such as carbon, chlorine, sulfur and iron didn't exist. The heavier elements such as tin, iodine and gold didn't exist either. This raises the question of the origin of many of the 92 stable elements essential to some planets and especially to our planet and the life it sustains.

Typical stars, such as our sun, do produce some of the medium weight elements up to iron. This is done through fusion when the temperature and density of the plasma in the cores of the stars are high enough. However, typical stars are not massive enough to create the conditions necessary to form elements heavier than iron.

Massive stars, eight to ten times the mass of our sun, can produce the heavier elements. They can produce all the rest of the elements needed. As indicated earlier, many of the first stars may have been supernovae as much as 300 times the mass of our sun. Supernovae also provide an excellent distribution system to spread the elements they produce, throughout a galaxy. See Figure 1.17.

After a long life, these massive stars die suddenly in a colossal explosion, which radiates with such brilliance that they are visible throughout the universe. That is why such exploding *stars* are known as *supernovae*. After supernovae have produced the heavier elements, they distribute them very efficiently when they explode.

Recently, astrophysicists have discovered some unusual properties of supernova explosions. They have found that some supernovae produce violent shock waves. It appears that these waves aid the distribution of the elements, by pushing huge jets of plasma far into the surrounding space…as far as hundreds of light years. The apparent abundance of stars, massive enough to become supernovae, is just one indication that there should be many solar systems capable of supporting life.

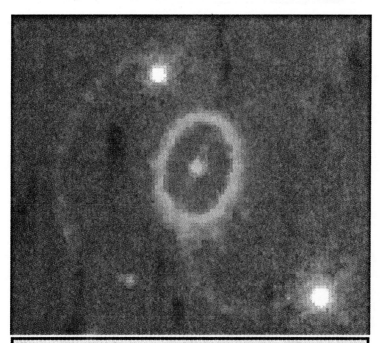

Figure 1.17 Afterglow of A Supernova. Supernova 1987A was observed, in 1987, radiating with a brilliance expected of these stars. The above photo, made in 1994, shows halos of hot gases, which are all that remains of the supernova. Recently, it has been hypothesized that supernovae as large as 300 times that of our sun formed very early in our universe, as early as a few hundred million years ATB. These supernovae are believed to have seeded the early galaxies with the heavier elements that smaller stars, like our sun, cannot produce.

OUR SOLAR SYSTEM

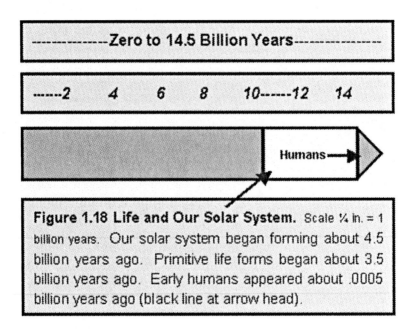

------------Zero to 14.5 Billion Years-----------

-----2   4   6   8   10------12   14

Humans→

Figure 1.18 Life and Our Solar System. Scale ¼ in. = 1 billion years. Our solar system began forming about 4.5 billion years ago. Primitive life forms began about 3.5 billion years ago. Early humans appeared about .0005 billion years ago (black line at arrow head).

Our solar system is located in the Milky Way galaxy and is about five billion years old. Our sun is an average star. It was formed in much the same way as trillions of other stars…by the slow and constant pull of gravity between the atoms of gasses and dust formed in the early universe. As described above, the general vicinity of many stars was sprinkled with atoms of the heavier elements from a supernova. This happened to our star, the sun. In addition, the sun produced some of the lighter and medium weight elements by fusion in its plasma core.

The age of our special planet, Earth, has been determined to be about 4.56 billion years. The dating was done by radioisotope dating of meteorites. The oldest meteorite tested is that old. It is generally believed that the Earth and other planets were formed from gasses and dust spinning in a disc and revolving around the sun fast enough to balance the pull of gravity.

This spinning disc of matter began to clump into blobs. The larger blobs dominated the merger process, thus forming the protoplanets. The gravity of each of the protoplanets held them together. Their mass and speed, along with their distance from the sun and the mass of the sun, determined the orbit of each planet around the sun.

As the protoplanets cooled, their gravity pulled them into spherical shapes. Solids and liquids, as well as gasses, initially formed on most of the planets. The structure of each planet was determined, to a great extent, by its mass and distance from the sun. In general, their atmospheres (if any), distance from the sun, and original temperature, determined the surface temperature of a planet.

The nine planets of our solar system, in order of their distance from the sun are: Mercury, Venus, Earth, Mars, Jupiter, Saturn, Uranus, Neptune and Pluto.

3

More Information From Space

The examples above are somewhat typical of what is going on in our universe. Information about such events is pouring in at an ever-increasing rate as more sophisticated telescopes, space probes and satellites are deployed. Let's look at some more examples of the mysteries and discoveries revealed by recent information from space.

QUASARS

Quasars were so named because they were thought to be quasi (somewhat like) stars. When discovered, by radio telescopes, it was believed that instead of radiating light, they radiated only radio waves. Later, it was realized that quasars are actually extremely brilliant sources of light of a broad range of frequencies, as well as radio frequencies. In addition, the light from quasars is red-shifted because of their great distances from the Earth.

Actually, quasars are not very much like typical stars. Recently, the Chandra X-Ray Telescope recorded a quasar that is 12 billion light years away. It is one of the largest and most distant objects observed to date. It was shinning when our universe was only two billion years old. Its radiation intensity is greater than that of a *trillion* stars! Quasars are *colossal* compared to typical stars.

Figure 1.19.The Hubble Space Telescope. This photo shows work being done on the HST in 2002. NICOMOS, a new infrared telescope, was installed. It can penetrate dust clouds far enough to produce images of the interior of star incubators. Figure 1.13 is a NICOMOS image of the tip of the Cone Nebula.

BLACK HOLES

The Hubble Space Telescope found that every galaxy examined in detail, also appears to have a black hole at its center. That is, of course, a very probable location for a large black hole, because it is the center of the colossal gravitational force of a galaxy.

When the density of a black hole reaches a critical point, its gravitational pull becomes so great that its existence can be *inferred* from its gravitational effect on gases swirling around the hole, much like water swirls down a drain. Measurement of the velocity of the swirling gases enables cosmologists to calculate how much compacted matter must be present in the black hole.

It's a lot. Typical black holes have a mass (weight) in the range of *four billion* suns like ours. In addition, such black holes are like voracious fiery garbage disposals. They consume gaseous matter at the rate of a million suns per year or more. *Anything* that reaches the *event horizon* of a black hole is sucked into the hole. See Figure 1.20. In addition, nothing, not even light, can escape the tremendous gravity of a black hole. (Recent theories suggest that some radiation and particles *may leak very slowly* from black holes.) It now appears that black holes are a major feature of quasars and galaxies and appear to play a major role in the evolution of our universe.

It should be noted, at this point, that black holes are encountered in superstring theories at the ultramicroscopic level of the particles of matter. Superstring-M theory provides surprising information about both colossal and ultramicroscopic black holes. We will revisit black holes in some detail in Section 3.

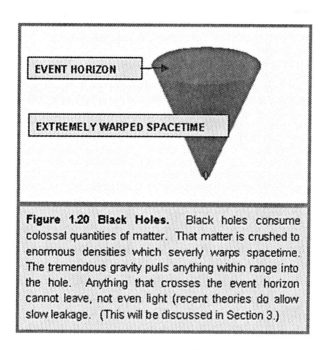

Figure 1.20 Black Holes. Black holes consume colossal quantities of matter. That matter is crushed to enormous densities which severly warps spacetime. The tremendous gravity pulls anything within range into the hole. Anything that crosses the event horizon cannot leave, not even light (recent theories do allow slow leakage. (This will be discussed in Section 3.)

THE EXPANSION OF OUR UNIVERSE

The big bang was hypothesized as a result of Edwin Hubble's discovery, in 1929, that galaxies were moving away from each other. That meant that the space (volume) of our universe was expanding. Some force was pushing or pulling space (and its components) outward. This discovery, along with relativity and quantum mechanics, led to development of the big bang theory of the origin of our universe. Calculations indicated that the force of a colossal explosion would cause rapid expansion. They also predicted that the expansion would *decrease* in time, because of the constant tug of gravity on the matter of our universe.

For years, cosmologists have been searching for evidence of a decrease in the expansion rate. One method utilizes measurement of the red shift in the spectrum of light emitted by supernovae. Although supernovae are dying stars, they are extremely bright when they flare; and they radiate light of uniform intensity. When they flare supernovae can be detected, as they existed as much as twelve billion years ago.

Imagine their surprise, a few years ago, when researchers found evidence (from the light of a supernova) that the expansion was *slower* seven billion years ago than it is now, fourteen billion years ATB. This is the opposite of the expected result; the expansion of our universe is *faster* now, not slower, as the universe grows larger and older.

What in our universe is going on? Well, the big bang theory could be wrong. On the other hand, there could be an unknown force at work. Such forces would have to be equal to, or greater than, the force of gravity and *repel* instead of attract matter. At the time Einstein developed the relativity theories, he was bothered that the stars in our universe were not moving toward each other. Telescopes, of that time, could not resolve certain stars (that were actually galaxies) into galaxies. The only galaxy known was our Milky Way.

Einstein realized that gravity should gradually pull stellar matter together. Since, at the time, that didn't seem to be happening, he added an ad hoc anti-gravity factor named *lambda* to his equations. Lambda became known as the cosmological constant. Einstein was uncomfortable about adding lambda because the only evidence for it was the lack of gravitational interaction between the stars. Later, he called lambda his greatest blunder. However, it now appears that Einstein's lambda *may* help explain the recent discovery about the expansion of our universe.

When expansion was discovered and explained by the big bang theory, lambda was ignored. It wasn't needed until recently. Now, lambda is being considered in a possible revision of cosmological theory. SS-M theory also will have something to offer on the problem.

OMEGA AND A MICROWAVE CONFIRMATION

Einstein's relativity theories indicate that matter curves (warps) the space around it. The greater the density of matter in a given volume of space, the greater the amount of curvature. Einstein's equations contain a factor, *omega*, which allows three possible geometric shapes for the space of our universe. If omega is *less than one*, our universe is open and saddle shaped. If omega is *more than one*, our universe is closed and spherical. If omega is *equal to one*, our universe is flat. See Figure 1.21.

A flat universe would not mean that the space around stars, planets and galaxies is not curved by their mass in their local volume of space. Rather, it would mean that the colossal volume of the universe, on *average*, is not curved. Remember, the colossal volume of our universe is mostly empty space.

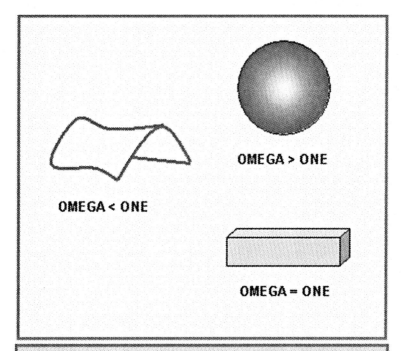

Figure 1.21 The Omega Possibilities. The curvature of the space of our universe is determined by the value of omega as shown above.

As long as fifty years ago cosmologists predicted that the big bang should have produced photons, and that they should have spread throughout our expanding universe. These photons couldn't have traveled outward until our baby universe became transparent to light, about 200 thousand years ATB. Since the temperature was still around 100 thousand degrees K at that time, the photons would have been very energetic.

However, during the 14 billion-year trip to the present, our universe's average temperature cooled to only 2 degrees K. During their trip the photons would have cooled too, thus reducing their frequency. Theory predicted that we should be able to observe them, in the present, as *microwaves* instead of light. Microwaves have a much lower frequency than light.

We can. A portion of the "snow" we see on a non-broadcasting television channel is produced by radiation from the big bang. The presence of microwave background radiation was confirmed experimentally in 1965; and the confirmation was reinforced by NASA COBE satellite experiments in the early 1990's.

The experimental data matched the predictions almost exactly. Results even indicate that the average number of photons is 400 per cubic centimeter. Obviously, these results show that the big bang theory is on the right track, at least from 300 thousand ATB. However, modifications will be necessary to resolve recent problems. Let's examine some of the problems and some possible solutions.

OMEGA AND THE CURVATURE OF SPACE

Needless to say, the microwave background radiation provides a window into the past and a very important source of information. Recent experiments used high altitude balloons with equipment designed to map predicted patterns of intensity. Since the microwaves have traveled over 14 billion years through space, they should indicate the value of omega.

Remember, the value of omega determines if and how the space of our universe is curved. Calculations of the sizes of patches of denser portions of our early universe were made from theory. This meant that if patches in the pattern appear *smaller* than predicted omega is *less than one*; and our universe is open and saddle shaped. If patches appear *larger* than predicted it means that omega is more than one; and our universe is closed and spherical. If patches appear *as predicted,* omega is *one*; and our universe is open and flat. These conclusions are based on the fact that electromagnetic waves are bent as they pass through space that is warped by matter (mass).

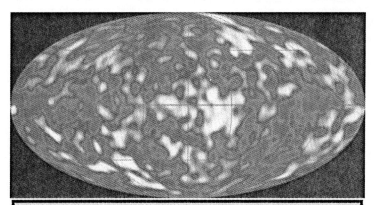

Figure 1.22 Messages from the Big Bang. This is a full sky map made from COBE satellite data. The pattern shows a slight temperature variation of only about one ten-thousandth of a degree. The darker areas are cooler by that much. The microwave background pattern is as predicted, and it indicates that omega is *one*. This means that the space of our universe is flat rather than spherical or saddle shaped.

Figure 1.22 shows one example of a microwave pattern from the COBE experiments. Surprisingly, the value of omega turned out to be *one*. Therefore, our universe is open and flat. Again, remember this doesn't mean that our universe is flat like a rectangular solid. It just means that the warping of space around stars, quasars, and galaxies, is not enough to curve the entire enormous volume of space into a sphere. It also means that the density (omega) is not small enough to curve space into a saddle shape. The density of matter, for a universe, is computed by dividing the enormous mass of all of its matter-energy by the even more enormous, mostly empty, volume of the universe.

An omega of *one* was unexpected. Inventories of the amount of matter in our universe yield, at most, only four percent of the amount of matter-energy needed to make omega equal to one. When estimates of recently hypothesized *dark matter* were added, the total increased to about 27 percent. That's still short by about 73 percent and, that is a big problem.

DARK MATTER-DARK ENERGY

At this point we have two problems: what force, or energy source, is speeding up the expansion of our universe; and where and what is the missing matter or energy? Well, dark matter was one of the first candidates for the missing matter. More recently, *dark energy* has been proposed as a source of a repulsive force. Perhaps, dark matter-dark energy, associated together as are matter-energy, is the answer.

As noted above, recent inventories, which include dark matter, have increased the amount from 4 percent to 27 percent of the omega one amount. That is still far short of the amount necessary to agree with the microwave pattern determination. Moreover, those experiments were made with state of the art monitoring equipment and were carefully verified.

Therefore, it appears, at this time, that omega is *one* and that space is flat, on average, within our universe. The missing matter-energy problem may be related to the fact that little is known about the properties of dark matter or dark energy. Although both concepts are in the speculative and investigative stage at this time, information is pouring in at a rapid rate. We will examine dark matter and dark energy in greater detail in Section 3.

THE DESTINY OF OUR UNIVERSE

At this time, two recent major discoveries (the increasing speed of expansion and the flat universe) are at odds with current theories. Both contradict conventional wisdom about the destiny of our universe. For a long time, it had been accepted by many that destiny would be determined by omega.

Now, it appears that dark energy, dark matter or some other source must be found to explain why omega is one. A closed universe (omega greater than one) was expected by many cosmologists. It was expected also that gravity would slow expansion to a stop and then contract our universe in a "big crunch", perhaps producing another nugget.

Then, possibly, the enormously compressed space and potential energy inside the new nugget would again produce a big bang; and the cycle could repeat to evolve another universe, and so on. Cycles do seem to be a favorite process of nature. We saw many examples of "gravity gathering" cycles in the formation of gaseous blobs, quasars, stars, galaxies, and planets.

In Section 3, we will discuss briefly a modification of the current inflation theory of the evolution of our universe recently proposed by Alan Guth. Soon, new developments may provide a satisfactory, verified solution to the question of the destiny of our universe.

Section 2

Superstring Theories

4

The Search for the Ultimate Particle

For centuries people have wondered about the stuff that makes up the things around them. Primitive societies considered such questions to be beyond them...as supernatural. They believed that unseen powerful beings caused thunder, lightning, wind and rainstorms. Moreover, they put the question of what things are made of in the same category. It is easy to understand why they believed that way. After all, they had very little accurate information about their environment.

UNCUTTABLE PARTICLES

Some two thousand years ago, Greek philosophers tackled the question with deep thought, observation and logic. They didn't put their theories to experimental tests, as we do today. They did, however, put their ideas through rather rigorous "mind experiments". On the question of matter, they concluded that if any sample of matter were dissected into smaller and smaller pieces one would eventually come upon a piece (particle) that is "uncuttable". They aptly named this particle the *atom*, which means uncuttable in Greek.

THE ELEMENTS

Centuries later, scientists discovered some ninety-two distinct kinds of stable elementary matter. These *elements,* as they are called, were thought to be the uncuttable particles of matter. Over the years the elements were studied in great detail. Groups of elements were found to have similar properties and were arranged into families in *The Periodic Table of the Elements.* In the following years, great advances were made in chemistry and physics because of this new knowledge of the elements.

ATOMS

Then came the discovery that the atoms of the elements are not truly elementary particles. Instead, the atoms are made of still smaller particles…*neutrons, protons* and *electrons*. Protons and neutrons form the nuclei of atoms. The electrons are whirling in orbits around the central nucleus. See Figure 2.1. This early description of atoms is the "solar system" analogy that, although not fully accurate, is still used in popular discussions about atoms.

The elements are still called "atoms" even though they are cuttable. Since the atom is the smallest particle of matter that identifies the elements, the name seems appropriate. Moreover, the atoms of the elements are the smallest particles that make up the molecules and crystals of "ordinary" matter.

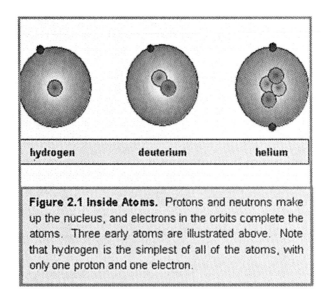

hydrogen   deuterium   helium

Figure 2.1 Inside Atoms. Protons and neutrons make up the nucleus, and electrons in the orbits complete the atoms. Three early atoms are illustrated above. Note that hydrogen is the simplest of all of the atoms, with only one proton and one electron.

OF QUARKS AND ELECTRONS

With the advent of "atom smashers", better described today as accelerator-colliders, a number of smaller particles were discovered. In 1968, in Stanford, CA when the linear accelerator was fired-up, the expected protons and neutrons were not found. Instead, they found *quarks* and other strange particles.

Then it was discovered that protons and neutrons are made of quarks. It was also determined that two kinds of quarks, *up-quarks* and *down-quarks*, make up neutrons and protons. Neutrons are made of one up-quark and two down-quarks. Protons are made of two up-quarks and one down-quark. See Figure 1.8.

At this time, it appears that electrons and quarks are indivisible. It also appears that all ordinary familiar matter, including all of the elements, is made of quarks and electrons. We must remember, however, that most of the matter of our universe is not the "ordinary" kind with which we are most familiar. Recent evidence indicates that perhaps only 4 percent of the matter-energy in our universe is ordinary. At this time, dark matter, dark energy and perhaps something else seem to be far more prevalent, totaling about ninety-six percent of all the matter-energy in our universe.

THE FUNDAMENTAL PARTICLES

By the end of the twentieth century the accelerator-colliders online were powerful enough to test the structure of matter to the infinitesimal size of one millionth of a billionth of a millimeter. As a result of many accelerator tests, most particle physicists are convinced that they have identified all of the uncuttable particles from which all ordinary and short-lived transitory matter is formed. (Numerous transitory particles such as pions and kaons are short-lived and are not a part of ordinary matter.)

The list includes the familiar electron, and two "cousins" of the electron, the muon and tau. There are three forms of neutrinos and six forms of quarks, making twelve fundamental particles. Examination of the properties of these twelve particles reveals certain family characteristics in three groups of four each.

The following table lists the particles and their masses grouped by families. The electric charges of the particles in each row are the same: Row 1: -1, Row 2: 0, Row 3: 2/3, Row 4: -1/3.

As we will see later, in both the supersymmetric standard theory and the superstring theories of matter, all twelve of these particles have (or have had in the past) an *antiparticle* and a *superpartner*.

There are several interesting comparisons between some of the particles. The electron, muon and tau all carry one negative electric charge. The mass of the tau is about 3,520 times that of the electron. The masses of the three types of neutrinos are believed to be near zero, but not quite zero. Experiments are under way to measure their masses. The masses of the six quarks vary from .0047 to 189. The top quark is about 40,200 times as heavy as the up-quark.

The standard theory provides no explanation for the apparent random values and wide range of the numbers. The unusual names of the quarks have no special meaning except to distinguish them from each other and perhaps, provide a bit of humor for the researchers.

Family 1		Family 2		Family 3	
Particle	Mass*	Particle	Mass*	Particle	Mass
Electron	.00054	Muon	.11	Tau	1.9
Electron Neutrino	$<10^{-8}$	Muon Neutrino	<.0003	Tau Neutrino	<.033
Up-Quark	.0047	Charm Quark	1.6	Top Quark	189
Down-Quark	.0074	Strange Quark	.16	Bottom Quark	5.2

Table 1. The Fundamental Particles
*The mass of a particle is compared to the mass of the proton, which is set at 1.0000.

THE FOUR FUNDAMENTAL FORCES

There are four fundamental forces associated with the fundamental particles. The multitude of interactions that occur between particles, as well as between large objects, can be reduced to only four fundamental forces. They are the *gravitational, electromagnetic, strong,* and *weak* forces.

The Gravitational Force

The gravitational force is very familiar to us. Gravity permeates our environment. We feel it. Its pull is measured by the weight of objects. Our bodies react to the force of gravity when we sit, stand, walk or do anything physical.

Gravity balanced by motion holds the planets and stars in their relative positions. The amount of matter (mass) in an object determines how much attraction it exerts on other objects. The distance between them and their masses, determine the forces between two or more objects. In "normal" space the gravitational force between objects decreases inversely as the square of the distance between them. For example, if a space ship moves two times as far away from the earth, it would then experience only ¼ the pull of the earth (2 squared=4, inverted=¼). The force particle of gravity is believed to be the *graviton*. However, the graviton has not, as yet, been detected experimentally. The search for the graviton is difficult because of the *intrinsic* weakness of gravity.

Although gravity plays a major role in our universe, it is a weak force at the particle level. However, the number of particles in an object multiplied by the weak gravity of each particle can yield a colossal total gravitational force. Thus, the greater the mass of an object the greater is its gravitational force. This is why gravity in stars becomes so colossal that fusion is ignited. An even more colossal force of gravity forms black holes. Therefore, we see that gravity can be considered to be either weak or strong, depending on whether we are referring to a few particles or a large group of many particles.

The Electromagnetic Force

We are not aware of the electromagnetic force in the constant physical way that we feel gravity. We are immersed in gravity and are directly aware of its force of attraction on our bodies and our environment. We are not as physically aware of the effects of the electromagnetic force. However, in a less physical way, we are aware of the force through nature's brilliant displays of sunlight and lightning.

More indirectly, we are certainly aware of the existence of the electromagnetic force through the many electric devices that have become so important in our lives. They are all expressions of the electromagnetic force…a force many times stronger than that of gravity. In fact, it is 10 to the 42nd power (10^{42}) times as strong!

Can that possibly be right? Yes, really. Remember, we are comparing *intrinsic* strengths. This means that we are comparing the force between particles. For example, an electron and a proton attract each other with an electromagnetic force much greater than the force of gravity between one electron and one proton. The electromagnetic force between charged particles is *intrinsically* strong.

Gravity seems stronger to a great extent because the electromagnetic force is *balanced* in ordinary matter by equal numbers of positive and negative particles. Unless the balance is upset, there is no way we can be aware of the force. The equal numbers of positive and negative electric charges effectively cancel each other. We can, however, upset that balance.

For example, you upset the balance when you walk across a carpet and get an electrical shock, as you touch a metallic object. The friction of your shoes sweeps-up extra electrons to your body, giving it an electrical charge. The electrical balance is upset. That is the basic way that both nature and people manipulate the electromagnetic force.

Force	Particle	Mass	Strength
Gravity	Graviton	0	1*
Weak	W and Z bosons	86, 97**	10^{39}
Electro-magnetic	Photon	0	10^{42}
Strong	Gluon	0	10^{44}

Table 2. The Fundamental Forces

*Strengths are compared to the strength of gravity *arbitrarily* set at 1.
**Weak force particles occur in two masses as W and Z bosons.

Unlike the electric force, there is no repulsive force of gravity, at least in our neighborhood, to cancel the gravitational attraction. That is another reason why ordinary everyday experiences with these forces are misleading with regard to their comparative intrinsic strengths

The Nuclear Forces

Unlike the gravitational and electromagnetic forces, the strong and weak forces are unfamiliar to most of us. This is understandable since they operate inside the nuclei of atoms, and the range of both forces is limited to subatomic distances. Thus, nuclear forces do not permeate our environment; and we do not directly experience them.

The Strong Force

The strong force binds quarks together to form protons and neutrons. It also binds one or more protons together along with neutrons to form the nuclei of atoms. This is no easy task since protons carry a positive charge and repel each other. It takes a strong "glue" to stick and keep them together. When discovered, the strong force seemed to be just the ticket; so the strong force particle was named the *gluon*.

The Weak Force

The weak force also operates inside the nuclei of atoms. It is involved in the process of decay of radioactive elements such as radium, cobalt-60 and uranium-235. The force particle of the weak force is the *weak gauge boson*. The weak force boson occurs in two varieties, *W* and *Z bosons*, which have masses of 86 and 97, respectively.

Both the weak and strong force particles work by reacting to the amount of *strong charge* or *weak charge* carried by protons, neutrons and other particles. This reaction is similar to that of the electric charge. However, the weak and strong charges are *not* the same as the more familiar electric charges.

Electrons, muons, taus, the six quarks and the three kinds of neutrinos all carry a weak charge. Neutrinos carry only the weak charge, have *almost* no mass, no electric charge and travel at near the speed of light. This has made it extremely difficult to learn much about them. Only recently, very rare weak-force interactions between neutrinos and other particles have been observed. We will discuss the fascinating neutrino in more detail in Section 3.

THE STANDARD THEORY

Theoretical and experimental physicists worked throughout the twentieth century to develop and refine the information summarized in Tables 1 and 2. As these particles of matter and their associated force particles were discovered and studied, various theories of their interaction were proposed and tested. The standard theory (standard model) evolved out of this research.

During this time, many physicists focused their attention on quantum mechanics. As a result, data from the many experiments related to quantum mechanics provided much of the input to the evolving standard theory of matter. The theories of relativity were pushed into the background while the focus was on quantum theory.

Although standard theory incorporates special relativity, no way has been found to include general relativity. Thus, the gravitational force is not explained in the standard theory. However, since gravity is so weak, compared to the other three forces; it wasn't a big problem at the particle level. Nevertheless, a bothersome problem did arise at the ultramicroscopic level, where the laws of quantum mechanics reign.

The Problem of Incompatibility

Thus, for decades physicists have had to use two theories: relativity for stars, galaxies, galactic clusters and large distances; and another, quantum theory, for subatomic particles and ultramicroscopic distances. Used separately, the two theories have produced remarkable agreement with experimental data. Nevertheless, two theories for one universe…a colossal universe built of infinitesimal particles? Surely, that can't be. Moreover, it's certainly not intellectually satisfying.

From the beginning, Einstein was troubled by this inconsistency. He spent the last thirty years of his life attempting to unify relativity with quantum mechanics. During the last few decades physicists have intensified their search for a unified theory. Finally, as we will see soon, a theory was developed in the last third of the twentieth century that includes gravity, as well as a resolution of the quantum problems.

It is remarkable that Einstein was on the right track. He just didn't have the experimental information, mathematics and a key idea necessary to solve the problem. Perhaps the best measure of his genius was his ability to theorize ahead of experimental data. That kind of thinking is increasingly important to theoretical physicists today.

5

String Theory

A BIT OF HISTORY AND A DISCOVERY

Like most viable theories, string theory has evolved through a number of versions since 1968. In that year a young physicist, Gabriele Veneziano, came across a purely mathematical formula developed 200 years earlier by a Swiss mathematician, Leonard Euler. Veneziano applied the Euler beta-function to problems of the strong force. The results were both dramatic and puzzling. The formula provided mathematical descriptions of many of the properties of the strong force. Physicists were intrigued because the physics underlying the formula was unknown.

In 1970, physicists Nambu, Nielsen and Susskind announced the results of their studies of the Euler beta-function. They demonstrated that if the protons and neutrons in the nucleus are assumed to be extremely tiny one-dimensional strings, Euler's function describes the nuclear forces exactly.

This was the first suggestion that point particles might be replaced by vibrating strings. It was also realized that if the strings were small enough they would still look like point particles at the experimental level of accelerator-collider technology of that time.

Within a few years after the string theory of the strong force was introduced, new higher-energy experiments indicated that some of the predictions of the theory were incorrect. Meanwhile, another approach, the quantum chromodynamics field theory also under development, met with such success that most physicists soon forgot string theory.

Early String Theory

A few particle physicists, however, were so impressed with the simplicity and beauty of the idea of replacing point particles with tiny vibrating strings, that they began a thorough investigation of the theory. They were surprised to find that the theory not only allowed vibrating strings that yielded gluons, but it also produced many other string patterns that had nothing to do with the strong force.

Then it was realized that the string concept might have been applied too narrowly when it was tried with the strong force only. After examining some of the strings that were emerging from the formula, it was realized that the properties of one of the strings matched exactly those of the hypothetical graviton.

Could vibrating strings be the source of *all* particles and forces? Physicists Scherk and Schwarz were convinced that was the case. In 1974 they announced that string theory not only explains the strong force; but it also is a quantum theory that includes gravity.

The announcement didn't make the splash one would have expected. However, as is the practice in science, the process of verifying started; and soon conflicts were found between string theory and quantum mechanics. Then in 1984, Green and Schwarz resolved the conflicts and also showed that strings could include all four forces. Soon, the process of building string theory resumed in earnest.

The First Superstring Revolution

Beginning in 1984 and continuing through 1986 hundreds of physicists throughout the world worked to develop string theories. Their combined efforts demonstrated that many features of the standard theory, including quantum mechanics, *emerge naturally from string theory.*

In addition, they were impressed by the simplicity and elegance of the concept of vibrating strings at the heart of all matter and forces. Many physicists felt this so strongly that they developed a deep devotion to the theory. This devotion set a number of physicists hot on the trail of string theory. This period of intense activity became known as the "first superstring revolution".

Soon it became apparent that although the idea is simple, there is a problem in the details. The mathematics proved to be extremely difficult and tedious. Even determining essential equations seemed to be out of the question. When this happens, physicists turn to approximation techniques. However, in string theory approximate answers to approximate equations gave unsatisfactory solutions to some values crucial to progress.

As a result, for almost a decade a few dedicated physicists struggled with the mathematical problems, making a contribution now and then. Most of the others turned to other projects. Some new idea was badly needed to revive the kind of effort essential to the continued development of string theory.

The Second Superstring Revolution

Finally, in Santa Barbara, CA in the spring of 1995, Dr. Edward Witten lit a fire under an audience of the world's leading physicists. He proposed a new approach to the mathematical difficulties of string theory. That set off the "second superstring revolution" that continues to this day. As the work on strings resumed, progress increased; and confidence in string theory grew.

It soon became clear that while the standard theory provides a framework built on the discoveries of the last century, it couldn't explain *how* the particles are produced or *why* they have the properties observed. In addition, the standard theory does not have an overall theoretical structure from which the particles and their properties emerge *naturally*. String theory does have such a structure, as we saw earlier. Now that we have made a strong case for string theory, let's turn our attention to some essential details of the theory.

KEY PROPERTIES OF STRINGS

We have seen that string theory has a theoretical superstructure. The only input the theory requires is one number, which will be introduced later. (In our discussions, "string theory" is used as the generic name for the several viable string theories.) All of the properties of both the ultra-small and the ultra-large should be explainable through string theory. Now, let's examine the key properties that make it all work.

The strings of string theories are not quite like the strings of spun metal or fibers that we are familiar with in everyday life. However, ordinary strings do exhibit most of the properties of the strings of string theory. In fact, you can demonstrate some of these properties for yourself.

Just fasten one end of a piece of string about twenty inches long to something. Pull the other end to make the string taunt; and pluck the string to make it vibrate, producing a sound. Hold the string so that the length is shorter. Pull the string about as taunt as it was before, and pluck the string. Notice the difference in the tone of the sound. Now, pluck the string as you change the tension (pull). You should notice the relationship between the length and tension of the string and the tones produced.

If you have musical talent, you probably have a hit song by now. In any case, you have the basics of a hit theory. Seriously, music does utilize some of the same concepts as string theory.

The Fundamental Particle

Although the physics of string theory is similar to the physics of music there are differences, some very strange. The strings of string theory that produce the particles of matter and the forces have only one dimension...*length*. All matter and force strings are exactly the same. The *length* and *tension* are the same in all of the strings in both matter and force particles. In addition, all of these strings are formed into *loops*.

Theoretically, the loops can vibrate in an infinite number of resonance patterns. See Figure2.2. Thus, *the string is the fundamental particle* of matter and the forces. (For now anyway.) Each primary particle of matter (quark, electron) and each force particle is produced by a unique resonance pattern of vibration. Vibration produces the properties of the particle, such as the mass.

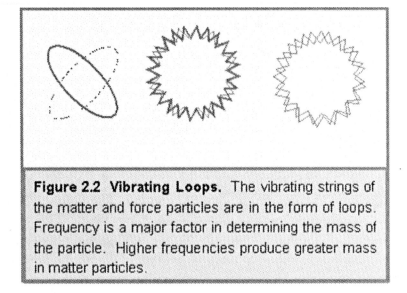

Figure 2.2 Vibrating Loops. The vibrating strings of the matter and force particles are in the form of loops. Frequency is a major factor in determining the mass of the particle. Higher frequencies produce greater mass in matter particles.

The above description of the basic premise of string theory is simplified. For instance, the strings not only vibrate, they also *spin* around an axis of the loop like an ice skater. They perform quite a dance. Spin takes care of certain properties, such as charge. Open-ended (non-loop) strings are also allowed in string theory. In addition, there are a few other properties that will be introduced as needed.

What Are Strings Made of?

So far, at least one important question has been ignored. Perhaps you have been wondering, "What are the one-dimensional strings made of?" They can't be made of any kind of matter, because matter is three-dimensional. Perhaps, they are made of *nothing*, or almost nothing...empty space. Earlier, strings were described as one-dimensional "slivers" of space.

Believe it or not, serious theoretical studies along these lines are under way. These studies involve efforts to determine the *intrinsic* properties of space and time. Some of the ideas include "imaginary time", and the nature of the extra dimensions of space, both curled-up and extended. Although the ideas are rather bizarre, they are no more bizarre than some of the conclusions of quantum theory, which we will be discussing next. (We will come back to string theory after the quantum material.)

6

A Bit of Quantum Theory

Quantum theory (also known as quantum mechanics) is rather difficult to explain fully, without the use of high-level mathematics. Some of the brightest physicists have said that even they don't understand the theory at a visceral level. However, we will start with some basic material and then explain quantum concepts as the need arises. You will enjoy learning about some of the bizarre effects of quantum mechanics.

Quantum theory came about around the end of the nineteenth century. The twentieth century was filled with quantum research, and it is still a major factor in particle physics research. Both experimental and theoretical physicists were kept busy investigating phenomena such as atomic structure, electrons and electricity, light and radioactivity.

A major development during the last half of the nineteenth century was Maxwell's electromagnetic field theory. This theory united electricity and magnetism, and it became the foundation for the technology behind the thousands of electric devices of today. Out of Maxwell's theory came the realization that light is an *electromagnetic wave*. The other forms of radiation also were determined to be electromagnetic waves, differing from each other only in wavelength (or frequency) and the amount of energy they carry. These new understandings provided many new challenges for the physicists of the time.

THE OVEN PARADOX

One especially bothersome challenge was, of all things, the paradox of a hot oven. Physicists were trying to calculate the amount of energy carried by the electromagnetic waves in an oven of a given temperature.

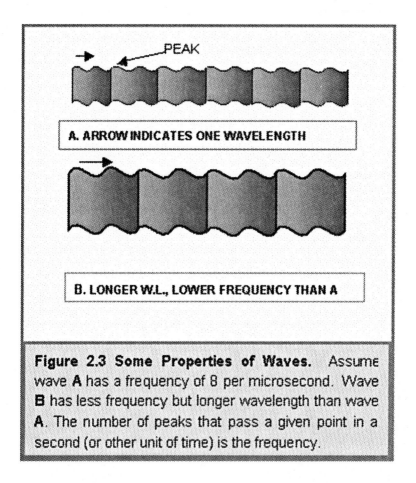

PEAK

A. ARROW INDICATES ONE WAVELENGTH

B. LONGER W.L., LOWER FREQUENCY THAN A

Figure 2.3 Some Properties of Waves. Assume wave **A** has a frequency of 8 per microsecond. Wave **B** has less frequency but longer wavelength than wave **A**. The number of peaks that pass a given point in a second (or other unit of time) is the frequency.

The sealed oven had a vacuum inside to eliminate convection heating. Calculations using equations of proven mathematics produced an absurd answer...an *infinite* amount of energy. This was nonsense and an indication that something was wrong. However, there was nothing wrong with the mathematics or the input data. Physicists were puzzled for quite some time.

Then, a physicist by the name of Max Planck, a name soon to be enshrined in quantum theory, applied Maxwell's electromagnetic theory to the problem. See Figure 2.3 for some properties of waves. Maxwell's theory established that only waves with a whole number of crests, or troughs, could function in the oven. However, since there are an infinite number of whole numbers, an infinite number of waves are still possible. In addition, it was thought that all waves, regardless of wavelength, carried the same amount of energy. Moreover, an infinite number of waves, carrying any amount of energy greater than zero, would still equal an infinite amount of energy. Of course, that can't be.

Well, Planck didn't give up. He proposed that the energy carried by electromagnetic waves comes in tiny *packets*. Furthermore, a packet was defined as the smallest unit of energy in a specific wavelength. In addition, the packets are not divisible. In other words, energy only comes in *whole number multiples* of a packet.

Carrying the idea further, Planck suggested that the minimum energy a wave can have is *proportional to its frequency:* higher frequencies (shorter wavelengths) have a higher minimum energy; lower frequencies (longer wavelengths) have a lower minimum energy. That did it. Not only did the new calculations yield a finite result, but the result was also in close agreement with the experimental data. The proportional factor that Planck used to resolve the oven paradox has become a major constant of quantum theory.

PLANCK'S CONSTANT

Yes, the proportionality factor determined by Planck is the well-known *Planck's constant*. It is an infinitesimal quantity…about a billionth of a billionth of a billionth of everyday units. The extremely tiny size of the constant means that the amount of energy in a wave packet is very small.

Planck's introduction of the quantum (packet) of energy set the stage for a whole new set of rules for dealing with the ultra-small: *quantum theory*. Although it resolved the oven paradox, at the time Planck didn't know how his idea worked. He introduced the key idea that the energy of electromagnetic waves is bundled into tiny packets, but he did not explain *how* the energy is bundled into the waves.

THE PHOTOELECTRIC PUZZLE

Then, in 1905, Einstein figured it out. He had been puzzling over a strange photoelectric phenomenon and realized that the energy packets might be the answer. The photoelectric effect (See Figure 2.4) occurs when light strikes the surface of some metals and agitates the loosely held electrons. Some of the electrons are kicked clean off the surface. (The electrons in the outer regions of metal atoms are loosely bound to the nucleus. This is the main reason metals are good conductors of electricity.)

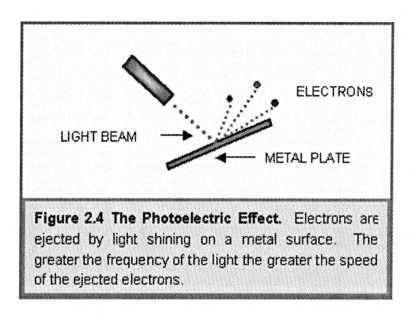

Figure 2.4 The Photoelectric Effect. Electrons are ejected by light shining on a metal surface. The greater the frequency of the light the greater the speed of the ejected electrons.

Intuitively, we would think that the greater the *intensity* (brightness) of the light, the faster the speed of the ejected electrons would be. Not so. Instead, as the intensity of the light is increased, the *number of electrons ejected* is increased.

However, experiments also show that that *the greater the frequency* (shorter wavelength) of light, *the greater the speed (and energy)* of the ejected electrons. For the visible spectrum, this means that red light has the least kick; and indigo light has the most kick.

Another interesting finding from these experiments was that when the frequency of the light is decreased, (wavelength longer) a point is reached when electrons are no longer emitted at all. This is because the packet does not have the minimal energy necessary to eject electrons.

WAVE-PARTICLE DUALITY

As Einstein pondered the photoelectric data and Planck's packets of energy with their relationship to frequency, he realized that light must consist of something in addition to waves. Finally, he declared that light consists of streams of tiny packets (particles) of energy associated with electromagnetic waves. Later, these packets, which are quanta of energy, were named *photons*. He further stated that the energy of a photon is proportional to the frequency of the associated light wave. The proportionality factor was, of course, Planck's constant.

Moreover, the connection between frequency and energy explained the connection between frequency and speed of electron ejection. The more energy a photon has, the greater the energy it imparts to the ejected electron. Einstein concluded that all electromagnetic waves have photons associated with them.

Einstein's proposal also provided the mechanism that explained how Planck's hypothesis worked. Any device that produces radiant energy, such as an oven, is limited in the amount of energy it can produce. The higher frequency waves, which require a minimum energy packet greater than the oven can produce, will not be formed in the oven. Thus, only a finite number of waves will be created to produce the oven's total energy. Again, Einstein's remarkable insight gave physics a major boost. We might say he made a quantum leap for physics. Or, maybe not.

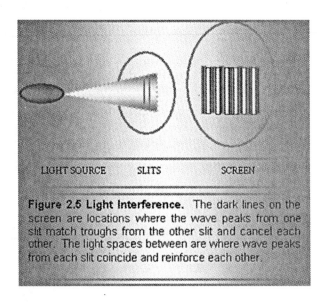

LIGHT SOURCE   SLITS   SCREEN

Figure 2.5 Light Interference. The dark lines on the screen are locations where the wave peaks from one slit match troughs from the other slit and cancel each other. The light spaces between are where wave peaks from each slit coincide and reinforce each other.

PARTICLE-WAVE DUALITY

If electromagnetic waves have particle-like properties, then perhaps matter particles have wave-like properties. Louis de Broglie was thinking that way, in 1923, when he proposed that matter particles, such as electrons, do have wave-like properties. Einstein had already proved the relationship between matter and energy: (E=mc²). Moreover, Planck's constant defined a close relationship between wave frequency and energy. The mathematics of particle-wave duality was worked out by de Broglie, and before long, physicists at Bell Telephone Laboratories found experimental proof of particle-wave duality.

The experiment consisted of projecting a stream of electrons onto the surface of a thin slice of crystalline nickel. A phosphorescent screen was placed on the other side to receive the electrons that pass through. Microscopically, the arrangement of the crystals in the nickel provides slits like those used to demonstrate light wave interference. See Figure 2.5. The electrons, which pass through the slits, are recorded by phosphorescence at the point where they strike the screen.

Well, as you probably guessed, the signature interference pattern of waves appeared on the screen, similar to the one shown for light in Figure 2.5. This means that electrons do exhibit the properties of both particles and waves. Before long, other particles such as the proton were tested. The conclusion was that all matter particles do appear to have particle-wave duality. Computation of the wavelengths of particle waves showed them to be extremely small. The wavelength was determined to be proportional to Planck's constant and the momentum (velocity X mass) of the particle.

All of the above left little doubt about particle-wave duality, except for one problem. Experiments were done shooting only one electron at a time several seconds apart, at the nickel target. Guess what. The interference pattern still appeared on the screen. Now, just how can particle-waves, separated by time, interfere with each other? Get ready for more weirdness.

PROBABILITY WAVES

In 1926, Max Born proposed that matter waves should be thought of as *probability* waves. He stated that the *square of the magnitude (size)* of a wave indicates the *probability* of where the particle is located (and where it hits the screen). The rule is: *locations where the square of the magnitude is higher have a higher probability that the particle will hit there.* Conversely, *places where the square of the magnitude is lower have less probability the particle will hit there.* Thus, what looks like an interference pattern of light and dark bands is formed on the screen behind the slice of crystalline nickel.

The probability wave also explains why electrons fired at high speed at a solid target are not shattered when they hit. Instead of shattering, the electrons simply appear at one of their probability locations. Obviously, if probability is operating at this sub-atomic scale, a strong element of uncertainty is operating throughout our universe.

Despite how strange all of this seemed, physicists tackled the job of checking it out. An equation was developed that describes the shape and evolution of probability waves. Then the equation was used to make probability predictions. Experiments proved the predictions to be very accurate. No exceptions have been found to this day. As a result, most physicists accept matter probability waves.

Classical theories, with their clockwork predictability, have been replaced with theories that have probability built into them. Our universe is still subject to a precise and strict mathematical framework. However, now it is assumed that only the *probability* of future events can be predicted. Although no one knows the fundamental structure underlying quantum theory, the theory has been accepted widely since about 1928, because it is in *astonishing* agreement with the experimental data.

THE "SUM OVER ALL PATHS" ALTERNATIVE

There is another theoretical approach to the problem of how electrons can cause the interference pattern, even though electrons are fired one at a time seconds apart. It is Feynman's "sum-over all paths" approach. See Figure 2.6. In this theory, a particle is considered as traveling over every possible path to the target. Probabilities are assigned to each possible path, and the average determines the probability that a particle (without a wave being involved) will arrive at a specific point. Strangely enough, or probably because quantum theory isn't fully understood, both theories give the same results and both are in use.

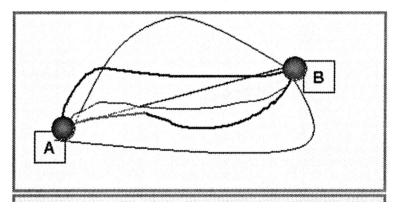

Figure 2.6 Sum Over All Paths. One explanation of how electrons fired one at a time can form an interference pattern assigns probabilities to each possible path a particle can take. The average determines the probability that a particle will hit a specific point. No wave is involved in this theory.

THE HEISENBERG UNCERTANTITY PRINCIPLE

In 1927, Werner Heisenberg proclaimed what has become known as the *Heisenberg uncertainty principle*. This principle states the problem of trying to experimentally determine, *simultaneously*, the position and velocity of a particle such as the electron. It had been realized for some time that probing with light (photons), or any other particle, disturbs the object being probed. For instance, when an electron is hit by a photon, *both the position and velocity of the electron are changed*. This indicates that we cannot know *both the position and velocity of a particle with precision*. This is because the *measurement process* changes both the position and velocity.

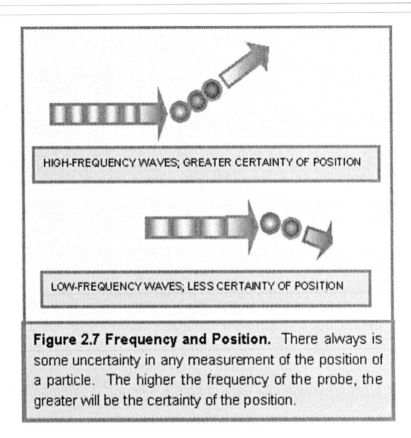

Figure 2.7 Frequency and Position. There always is some uncertainty in any measurement of the position of a particle. The higher the frequency of the probe, the greater will be the certainty of the position.

One might argue intuitively, that if we reduce the energy (frequency) of the light and use super-sensitive light detectors, the effect on the particle will be negligible. All of this would help, but Heisenberg's point was that there is a limit. The limit is the wavelength of the light. Probes of longer wavelength yield less precision. Therefore, if we reduce the frequency, as suggested above, the longer wavelength (lower frequency) decreases the precision. It seems we can't win for losing.

We can, however, determine *either* the position or velocity with the *highest possible precision*. That is, we use higher frequency probes to determine a particle's position. Conversely, we use lower frequencies to determine velocity. The above relationships are illustrated in Figures 2.7 and 2.8.

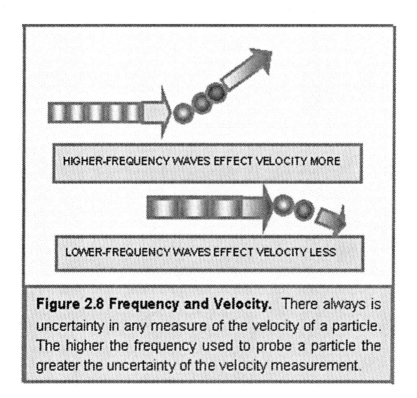

Figure 2.8 Frequency and Velocity. There always is uncertainty in any measure of the velocity of a particle. The higher the frequency used to probe a particle the greater the uncertainty of the velocity measurement.

Heisenberg stated the mathematical relationship between position and velocity measurements to be: *the greater the precision of position measurement, the less the precision of velocity measurement.*

Heisenberg also determined that the precision of energy measurements is related to the amount of *time* required to make the measurement. In other words, *the precise energy of a particle cannot be determined for a precise moment in time.* Again, the *greater the precision* of the energy measurement, the *longer the interval of time* needed to make the measurement. See Figure 2.9. He also found that all of the above rules are true for any matter particle and any equipment used.

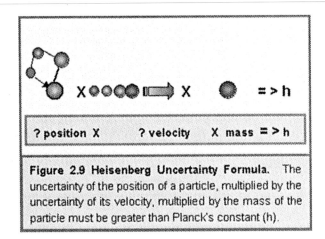

? position X   ? velocity   X mass = > h

Figure 2.9 Heisenberg Uncertainty Formula. The uncertainty of the position of a particle, multiplied by the uncertainty of its velocity, multiplied by the mass of the particle must be greater than Planck's constant (h).

For many years some of us, including Einstein, argued that even though we cannot know, simultaneously, both the precise position and velocity of a particle, it still *has, simultaneously,* a definite position and velocity. Well, it is now being asserted that recent experimental results prove us wrong.

QUANTUM TUNNELING

Sometimes they do it in movies, but most people don't even try it. Walking through solid walls is not considered possible by most of us. However, quantum mechanics does indicate that that there is a *small* probability that it can be done. The probability is so low that it would happen only once in about 15 billion years. Once, that is, if you tried it once every second, non-stop, for 15 billion years. If you want to do the experiment you will need a lot of patience and a "many generations" relay team.

Particles, such as electrons, have a much better chance of penetrating the wall. From the viewpoint of an electron, a solid wall is an energy barrier. Believe it or not, the electron can *borrow* enough energy to penetrate the barrier for a *moment.* The energy can be borrowed only for a moment, because the borrowed energy must be paid back, very quickly.

This process is known as *quantum tunneling.* It works a little like a checking account. As you probably know, it is possible to pay a bill with a check, before the money is deposited to cover the check, without your account being overdrawn. Of course, sufficient money must be deposited and credited, *before* the bank's accounting system deducts the check, or your check will bounce.

This is somewhat like the electron borrowing energy from the universe, and then paying it back before the universe can react. Anyway, that's a little like what particles do when they go quantum tunneling. This is possible, according to quantum theory, as long as the time interval of the process is within a set limit. The limit is determined by the Heisenberg uncertainty principle and is about 10^{-21} of a second for a typical positron-electron appearance and disappearance. In other words, *the particle must pay back the borrowed energy within that time limit.*

QUANTUM FRENZY

The uncertainty principle tells us that at Planck-size scales and Heisenberg time intervals; our universe is a very strange place, indeed. Even in the vacuum of "empty" space, uncertainty indicates that wild, frenetic, constantly changing activity is taking place. Moreover, as the space and time become smaller and smaller, the frenzy becomes greater and greater. Let's see if we can clarify this strange phenomenon with a thought experiment.

Imagine an electron in the middle of a tiny box. The box contains nothing but empty space (a vacuum) and an electron. The box has movable walls, like in a horror movie. As the walls slowly move inward, the electron becomes agitated. As the electron's space decreases, the electron becomes more and more energetic. If we could look into the ultramicroscopic space, we would become aware of an extremely chaotic situation as the electron very rapidly borrows and repays energy, keeping its frenetic dance going.

Heisenberg uncertainty maintains that in empty space, as in the above box, even without the electron or any other particle, there is a constant frenetic change in energy levels. The energy within the box fluctuates between extremes, and as the space gets smaller the extremes get larger. The space inside the box is constantly borrowing energy from our universe and quickly paying it back.

If the fluctuating energy extremes become large enough, the energy can convert into matter because Einstein says so: $E=mc^2$. The energy might, for example, convert to an electron-positron pair (matter-antimatter). In this case, they would "instantly" annihilate each other, converting their mass back into energy, thus repaying the borrowed energy. Quite a clever scheme.

There are many other kinds of matter-energy interactions that can occur in "empty" space. It would seem that, although empty space appears to be empty, it actually has, for extremely short time intervals, alternately positive energy and negative energy which *average to equal zero*, or near zero. At our everyday size and time scales, space appears empty. Moreover, at our current technological level it tests to be empty.

There is, however, an indirect experiment, which indicates that something is going on in "empty" space. The experiment demonstrates what is known as the "Casimir effect". As illustrated in Figure 2.10, when two parallel metal plates are placed close together in a vacuum, they move still closer to each other.

This is explained by the fact that the number of wavelengths that can fit between the plates is less than the number of wavelengths that exist outside the plates. This is interpreted as a reduction of the energy density (more accurately the energy density of ground state fluctuations) between the plates. The greater energy density outside the plates pushes them a bit closer together. This effect has been detected experimentally, and it appears to confirm the existence of ground state energy fluctuations in "empty" space.

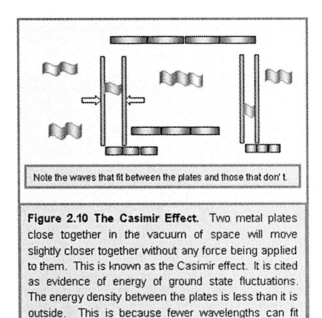

Note the waves that fit between the plates and those that don't.

Figure 2.10 The Casimir Effect. Two metal plates close together in the vacuum of space will move slightly closer together without any force being applied to them. This is known as the Casimir effect. It is cited as evidence of energy of ground state fluctuations. The energy density between the plates is less than it is outside. This is because fewer wavelengths can fit between the plates than exist outside them. The greater energy outside pushes the plates a bit closer to each other, as shown on the right.

QUANTUM EXPLAINATIONS

In the 1930's and 1940's physicists worked diligently to explain this strange, chaotic behavior of empty space. Since matter-energy transformations are involved, they realized that matter-energy equivalence ($E=mc^2$) should be a part of any theory. They also realized that Edwin Schrodinger's quantum wave equation, developed to deal with wave-particle duality, indicated a way to approach the problem.

Out of their work came *quantum electrodynamics*. The theory incorporates special relativity, and it is referred to also as the relativistic quantum field theory. The theory may be thought of in terms of classical fields of invisible lines of force, with photons sprinkled along the lines of force. These lines of force are constantly shifting from one quantum field to another...vibrating in spacetime.

The quantum field theory has produced a mountain of both theoretical and experimental data. The agreement between the theoretical and experimental is extremely good. The mechanics of photon-electron interaction and the role of the photon in electromagnetic phenomena has been defined and affirmed by the theory of quantum electrodynamics. The fact that it works is demonstrated by the great variety of electronic devices at work in our lives everyday.

In the 1960's and 70's similar approaches were applied which produced field theories for the weak and strong forces. They are the *quantum electroweak theory* and the *theory of quantum strong dynamics*. You may recall, from Section 1, that all three of these forces, as well as gravity, condensed out of the chaotic hot plasma that spewed from the big bang. This assumes that at the extremes of density and temperature that existed at the time of the big bang, the four forces were united as one force.

THE PROBLEM OF INCOMPATIBILITY AGAIN

The standard theory has been very productive for many years. Physicists use general relativity when dealing with the large portion of our universe. They switch to quantum theory when dealing with the ultra small. During the last several decades, two types of additions have greatly enhanced the standard theory. They are supersymmetry and the quantum field theories of the weak, strong and electromagnetic forces.

Efforts to develop a quantum field theory of the other force, gravity, within the supersymmetric standard theory have not been successful. Calculations using equations that merge quantum and general relativity theory yield infinity. As we have seen before, this means that something is not right. It was soon realized that the two theories are incompatible. Moreover, it appeared that the problem was related to the concept of empty space and the point particles of the supersymmetric standard theory

According to Einstein, empty space, remote from any mass is flat, smooth and homogeneous. However, empty space, according to Planck and Heisenberg, is filled with frenetic fluctuating energy fields. Obviously, the two theories are incompatible at sub-Planck scales.

Acceptance of quantum field fluctuations is a major tenet of quantum theory. One consequence is that fluctuating gravitational fields must exist throughout space. Even where the force of gravity is zero, the fields must exist. (That clever trick at work, again.) This quantum description of empty space is in graphic conflict with Einstein's flat, smooth, homogeneous empty space. Moreover, there is the feeling that many of us have that the elegance, the complexity (yet simplicity) and the perceived intelligence evidenced in the structure of our universe, must be explainable by *one all-inclusive theory.*

We have seen the problems with the standard theory; and that even with the addition of supersymmetry, the problem of incompatibility is not resolved. It appears that we need a fundamental change…a new approach. Fortunately, there is a high probability that we are on the threshold of developing a theory that solves the problem. A theory that replaces the standard theory's point particles with miniscule loops of string…strings that vibrate in Planck-sized regions of the curled-up space of extra dimensions.

7

Back to String Theory

TENSION AND THE GRAVITON

As we learned earlier, in 1974 Schwarz and Scherk reported that one of the resonance patterns of strings exactly fits the hypothetical graviton. Since the graviton was the missing link in the standard theory, this intrigued physicists working on that problem. In addition, based on the string properties of the graviton, they were able to calculate the *tension* in the graviton string. Moreover, since all strings in string theory have the same tension, this provided the crucial *constant* needed in string equations.

As you know, the intrinsic strength of gravity is very weak. In addition, in the equation that determines the force of the graviton the tension constant is an inverse factor (divisor). This leads us to expect that the tension should be huge (a small value for the weak gravity would be produced by dividing by a very large number). It is more than huge. It is a colossal 10^{39} *tons!* That is a thousand, trillion, trillion, trillion tons!

THE LENGTH OF A STRING

Using the newly determined value for the tension, calculations were made to find the length of a typical string. This time, we can expect a very tiny number because the extreme tension should shrink the string loop to *almost* nothing. It does turn out to be extremely small, and then some…10^{-33} of a centimeter. This is also the value of the Planck length.

It would seem that is about as small as anything can be. However, it is not as small as the particles of standard theory, which are point particles of *zero* size. This difference, between the matter and force particles of the two theories, a miniscule string versus a zero point particle, is a major factor in resolving the conflict between general relatively and quantum theory.

THE ENERGY OF STRINGS

We all know by now, that higher tension in a string produces a higher frequency of vibration. We also know that higher frequencies produce greater energy. There is one more factor that determines the energy of a string. It is the *amplitude*, or height, of the wave…the *higher* the amplitude, the *greater* the energy of the string will be. Remember, the amplitude of a wave determines the number of packets of energy a wave can carry. In addition, the number must be a whole number.

Therefore, the three factors, which determine the resonant vibration patterns of strings, are: *tension* (which is constant), *frequency* and *amplitude*, both variables. The frequency determines the amount of energy in a packet, and the amplitude determines the number of packets. Together, along with the tension, they determine the total energy.

You may be wondering how all of the above translates into vibrating strings that correspond to the matter and force particles in Tables 1 and 2. Any viable theory must be in agreement with these tables. That is because the data in the tables is from the extremely well tested data of the standard theory. Well, the graviton popped up early in string theory. It is a plus for string theories, since the standard theory doesn't have a graviton. The colossal tension of strings means that even the minimum energy of strings is *very large*. This minimum energy is expressed in multiples of the Planck energy.

In order to get a better concept of the size of the Planck energy, we will use the Planck mass, which is the equivalent of the Planck energy (they can be converted, one to the other, by $E=mc^2$). *The Planck mass is 10^{19} greater than the mass of the proton.* In other words, the Planck mass is *10 million trillion* times the mass of the proton! This is staggering to the mind when you recall the ultra small scale in which all this is happening. We can see that string loops should produce *very massive* particles.

Does it seem like we are working ourselves into a corner? It does, because the key particles of string theory are nowhere near that massive. We need quarks, electrons, photons and about thirteen other particles. Moreover, an electron, for example, is only .00054 as massive as a proton. This appears to be an insurmountable problem. We have vibrating strings that produce a minimum mass that is almost 2 billion trillion times that of our heaviest fundamental parti-cle...the quark. (The heaviest quark is 189 times as heavy as the proton.)

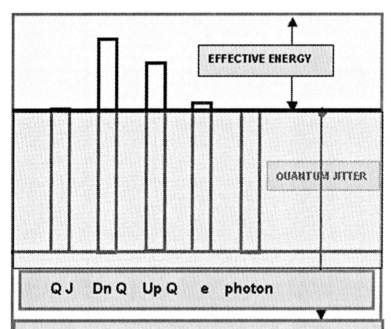

| Q J | Dn Q | Up Q | e | photon |

Figure 2.11 Frequency Energy and Mass. Shown above, left to right, are: quantum jitter, down-quark, up-quark, electron, and photon energies relative to each other. All of the particles have frequencies far greater than their effective energies (and masses) as indicated by the top arrow. If the vertical length of the energy bars in the graph were to scale they would extend trillions of inches below the gray portion of the graph. In each case quantum jitter cancels most or all of the energy (mass) created by the vibrating string.

Quantum mechanics to the rescue. Again, Heisenberg uncertainty assures us that all particles, including strings, constantly undergo quantum jitter. Yes, the loops of string undergo quantum jitter while they are also vibrating in their reso-nance patterns. But guess what. The energy of the quantum jitter is *negative,* and the energy of the resonant patterns is *positive*. Astoundingly, the negative jitter energy is roughly one Planck unit of energy (or mass). See the graph in Figure 2.11.

I'm sure you know the rest of the story. Yes, the negative jitter energy cancels about one Planck unit of energy of the resonant pattern energy of all vibrating loops. If a first level string produces one positive Planck energy; and if jitter produces one negative Planck energy, the mass of the particle is zero; and the particle could be the graviton.

The other matter and force particles of Tables 1 and 2 are produced in a similar manner. For example, if the electron resonance pattern produces 1.00054 Planck mass, and we subtract 1.00000 Planck mass for negative quantum jitter, we have the mass of the electron: .00054. Well, we got out of that corner very nicely. (I'm sure you're beginning to see why we stopped the study of strings a while back to learn a bit about quantum theory.)

WHERE ARE THE MASSIVE PARTICLES?

Sorry, we have another problem. The problem arises because strings can vibrate in an *infinite* number of resonance patterns. This means that theoretically, any whole number times the energy in an energy packet can be produced. Since there are an infinite number of whole numbers (1,2,3,4,etc.) our universe should be swarming with an infinite number of different kinds of particles. Or, maybe not.

All but the level one particles would be heavy particles. This means that particles from two to *many* times the Planck mass would be heavier and heavier, as the number increases. Note that the number of whole numbers used was stated as *many* instead of infinite. This time, the theoretical infinity could not be reached because the *amount of energy available* when the particles were created was not infinite. That would have been soon after the big bang, when there was a colossal amount of energy but not an infinite amount.

Still, if only a colossal number of heavy to massive particles should have been produced, where in the universe are they? The major reason we haven't detected them may be that they decayed, in steps, producing a cascade of lighter and lighter particles. (This would be similar to radioactive decay in some of the heavier atoms.) The decay would have ended at level one, near the Planck mass. Thus many, or most, of our fundamental particles may have been formed in this way. If the decay theory is correct, it probably explains what happened to most of the massive particles. However, a few might still be around.

The accelerator-colliders of today, don't achieve anywhere near the energy needed to produce the really massive particles. Therefore, we wouldn't expect to produce and detect such particles with the current colliders. There is some expectation, however, that by increasing their power and using special techniques, some of the larger colliders may be able to produce and detect some of the lighter heavy particles. This would be a major event and would lend great support to superstring theory.

Now, let's look at a more speculative hypothesis for what happened to the massive particles. Perhaps the massive particles came together and positioned themselves somewhere in our evolving universe. This probably would have happened early in their formation. Their gravity would have been so much greater than that of the fundamental particles that it is easy to see why they might have formed clusters early in the matter formation process.

In addition, their total energy-mass would have been huge, compared to that of the fundamental particles. It even appears that it would be enough to account for some of the missing matter-energy indicated by the confirmation that omega is one. Recall that for omega to be one, our universe appears to have less than four percent of the necessary matter-energy. It does seem that *if some massive particles exist* they could account for at least a part of the missing 96 percent of matter-energy.

The 96 percent is now being referred to as *dark matter* and *dark energy*. It is assumed that matter could exist that does not radiate detectable energy. If so, we could not observe dark matter with any of the many devices, such as telescopes, that use radiant energy to "see" ordinary matter. Recently, however, dark matter was detected indirectly, by its gravitational effect on galactic clusters. We will explain this and discuss dark matter and energy more thoroughly in Section 3.

Large concentrations of matter and energy could be hidden in black holes. Although black holes can be detected by their gravitational effect on nearby matter, they are often hard to find. Moreover, some could be in parts of our universe where there isn't enough matter left nearby to make them detectable. In fact, we don't know how many black holes exist in our own galaxy, much less in our universe. Efforts are underway, however, to conduct a search and census of black holes in our galaxy.

The massive particle hypotheses as a source of dark matter may not be valid. After all, most theoretical physicists believe the massive particles would decay into the fundamental particles. The massive form may be the optimal form for the particles to take *when initially formed* in the conditions soon after the big bang. Those conditions were extreme beyond imagination. It could well be that only massive strings could form under those conditions, and it was a necessary first step in forming the strings of the fundamental particles and forces. The final step would have been a cascade of decay into the fundamental particles.

WINDING STRINGS

Although not essential to the basic understanding of the matter and force particles of string theory, there is another function of strings that has been mentioned but not discussed. Some open strings wind around curled-up dimensions. See Figure 2.12. They are called *winding strings*. Winding strings have the same basic properties as the strings we have been discussing. However, there are differences.

For instance, they are open-ended and have a minimum mass. The distance around the encircled dimension, and the number of times the string winds around it, determine the minimum mass. In general, the longer the string is the greater its minimum mass will be. Any vibration of the string adds to its mass.

Wound strings are also referred to as *wrapped strings*. The dimensions are referred to as *wrapped dimensions*. The concept of wrapped dimensions leads to a possible explanation of why there are only three extended spatial dimensions.

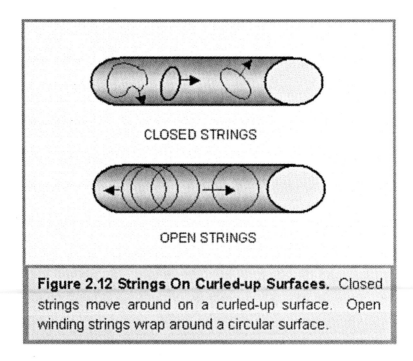

CLOSED STRINGS

OPEN STRINGS

Figure 2.12 Strings On Curled-up Surfaces. Closed strings move around on a curled-up surface. Open winding strings wrap around a circular surface.

A 3-D EXPLAINATION

Recall that, soon after the big bang, three of the compacted spatial dimensions unfurled and expanded to form the three-dimensional visible portion of our universe. You may have wondered why only three of the original compacted spatial dimensions unfurled. String theory provides an answer.

It appears that all of the compacted dimensions were wrapped with strings that kept them compacted. However, there is a clever mechanism that allows some wrapped strings to unwind. If a wrapped string and an *antistring* come into contact, they annihilate each other. This process produces an unwrapped string. (Strings and antistrings wind around dimensions in opposite directions.)

The *probability* that a string and an antistring will collide and produce an unwound string is *large enough only in one, two or three dimensions*…a maximum of three. Thus, as their wrapped strings tried to unwind, in the aftermath of the big bang, three of the compacted dimensions were freed from their bonds. They became the visible three-dimensional extended portion of our universe. The other curled-up dimensions remained wrapped and became the invisible ultramicroscopic portion of our universe.

SYMMETRY AND STRING THEORY

Symmetry in Nature

The laws of physics are expected to be true, no matter where or when they are operating. In other words, *nature treats every place and moment the same*. The fundamental laws apply everywhere and every time, *equally*. This is a principle of symmetry that appears in the laws of nature.

This does not mean that there are no differences in the *effects* of the laws in different places or moments. In fact, the effects are often drastically different. For example, the laws of gravity are the same on mars and on earth. However, the effect (pull) is much less on mars, because mars has less mass. Mass is one of the variable factors in the equations of the laws of gravity.

Symmetry and Relativity

A principle, which is crucial to special relativity, states that all physical laws must *stay the same* regardless of the constant velocity of the relative motion of different observers. This is another example of symmetry in the laws of nature because different observers with different constant velocities are treated alike. Again, as with gravity, the observations will reflect the different motion, but the *underlying law* will be the same.

Einstein expanded this symmetry by demonstrating that it is true for all observers…even those experiencing acceleration. This means that the force of gravity cannot be distinguished from the force of acceleration. This equivalence between the forces of gravity and acceleration is a major symmetry of relativity.

Rotational Symmetry

We discussed rotational symmetry in Section 1 in relation to phase changes that bring about symmetry breaking and building. We saw that both liquid water and steam exhibit rotational symmetry. This is because, when examined on the molecular scale, water looks the *same* regardless of the *rotation of the viewing angle*. Steam, also looks the same from any angle. This is an example of the democracy of physical laws, since all observers are treated equally, regardless of viewing angle.

The Final Symmetry: Quantum Spin

Note that the above symmetries deal with position, motion and time. By the late 1960's, after theoretical and experimental considerations, physicists decided there were no more symmetries of this type to be discovered. However, as early as 1925, the rotation (spin) of orbiting electrons was postulated as a source of magnetism that could explain the mechanics of the emission and absorption of light by electrons. This idea proved to work quite well and is still in use.

The spin of the electron was determined to be an *intrinsic* property of all electrons. Furthermore, *all electrons spin at the same constant never-changing rate*. Although this spin is somewhat similar to that of a spinning top or ball, it is a kind of *quantum spin*.

After checking the other matter particles, it was found that all twelve have the exact same spin as the electron. What's more, all of the twelve antimatter particles have the same spin. The spin of the matter particles was designated as ½, which is, roughly, a measure of the quantum mechanical speed of the spin. As you probably guessed, the force particles also have spin. The graviton has a spin of 2. The gluons, W and Z bosons and photons all have spins of 1.

It wasn't until 1971 that physicists realized that the subtle quantum spin of the particles was too unique to be considered as rotational symmetry. The derivation of this symmetry is highly mathematical, difficult to follow and beyond the scope of this book. If you take it with a grain of salt, your concept of a spinning top will serve our purposes quite well.

The new symmetry came to be called *supersymmetry*. In a way, it ranks above the other symmetries; and it certainly has lived up to its name. As indicated earlier, it has been incorporated into both the standard theory and string theories. These new versions of the theories are known respectively, as the *supersymmetric standard theory* and the *superstring theories*. As a result of supersymmetry, these theories make predictions that could experimentally prove or disprove supersymmetry as one of nature's laws in our universe.

SUPERPARTNERS

Not long after the discovery of supersymmetry, physicists realized that if included in their theories, supersymmetry would increase the number of fundamental particles again. Supersymmetry requires all matter and force particles to come in pairs. This is true for both the point particles of the supersymmetric standard theory, and the vibrating string loop particles of superstring theories.

The new hypothetical particles are called *superpartners* or *superparticles (sparticles)*. Superpartners differ from their partners only in their mass and spin. The spin of the superpartners must differ from their partners by ½ unit. For example, since the electron has a spin of ½, the selectron (prefix "s" for super) should have a spin of 0, and so on. At first, it was thought that some of the spin 1 force particles might be superpartners of some of the matter particles. However, that didn't check out.

Therefore, if supersymmetry is one of the symmetries of nature, there should be a whole new set of particles…a superpartner for each matter and force particle. As with the heavy mass particles, none have been detected. Again, one possible reason is that they are also too heavy to be produced by the power of the particle colliders that are currently in operation. However, the masses of the superparticles should not be as heavy as the massive particles discussed earlier. Again, since a number of the world's colliders are being upgraded there is a good probability that the lighter superpartners will be observed within a few years.

Anyway, it's beginning to seem like nature is playing a game with us; and we don't know all of the rules. That is probably the case; since quantum strangeness is involved. Supersymmetry does improve both the standard and string theories. In addition, supersymmetry has enabled theorists to develop an explanation for the strange, huge range in the strengths of the four forces. It turns out that the forces (gravity, electromagnetic, weak, and strong forces) could have been *unified* at the extreme temperature and density at the time of the big bang.

Furthermore, as we saw in Section 1, they would have condensed out of the plasma one by one, as the temperature dropped to the right level to separate each force. This success, and others, gives strong support for supersymmetry in our universe and in Superstring-M theory.

8

Curled-up Shapes and Extra Dimensions

EXTRA DIMENSIONS

As long ago as 1854, Georg Rieman introduced geometry of n-dimensions. Riemannian geometry can handle any number of dimensions. Unlike Euclidean geometry, Riemannian geometry can have space of four or more dimensions. This was exactly what Einstein needed for his theories of relativity. For one thing, he needed four dimensions in order to include time as a fourth dimension.

The three spatial dimensions of plane geometry, up, down, left, right, back and forth, are at right angles to each other. Thus they are independent; that is, they don't cross each other in a flat plane. Einstein's past-future dimension is obviously not a spatial dimension, but is one of time. However, time has importance equal to that of the spatial dimensions in specifying the location of an object or event.

For example, if you show up at the right street, street number and floor for a concert on the wrong day (time), you miss the concert. Although the duration (time) of an object or event can be short or long, they all exist for a specific amount of time. Riemannian geometry allowed the inclusion of time as the fourth dimension in the mathematics of relativity.

The subject of extra dimensions was a hot topic of debate and discussion in intellectual circles, during the latter nineteenth and early twentieth centuries. Writers and artists were especially intrigued by the possibilities of a multidimensional universe. The introduction of Einstein's special relativity in 1905, and general relativity in 1919, fueled interest and speculation about the meaning of other dimensions.

Picasso painted his interpretation of the effects of other dimensions on the appearance of objects. In addition, science fiction writers found a new realm to explore. From Carl Sagan's *Contact* to *Star Wars* and *Star Trek*, writers and producers have adopted cutting edge ideas of physics because they are interesting and exciting. Even some theologians claimed to have found Heaven in an extra dimension.

The three-dimensional nature of our world of experience is so embedded in our minds that the idea of more spatial dimensions seems ridiculous. However, encouraged by the success of the fourth dimension, a mathematician by the name of Theodore Kaluza was bold enough to challenge the strong belief that there are only three spatial dimensions. Kaluza declared that there may be more than three…perhaps, many more. Despite how bizarre this seemed at first, the suggestion would soon shake the foundations of physics. In addition, it supplied string physicists with an essential concept…*extra dimensions*.

Curled-up Dimensions

In 1926, another mathematician, Oskarl Klein, suggested that the extra spatial dimensions could be very small and curled-up. Later, Kaluza and Klein collaborated in combining and refining their ideas. They assumed a circular curled-up shape of one extra dimension and concluded that circular dimensions could exist at every point of extended three-dimensional space.

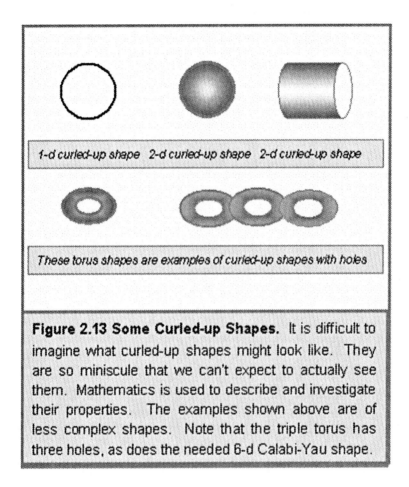

1-d curled-up shape   2-d curled-up shape   2-d curled-up shape

These torus shapes are examples of curled-up shapes with holes

Figure 2.13 Some Curled-up Shapes. It is difficult to imagine what curled-up shapes might look like. They are so miniscule that we can't expect to actually see them. Mathematics is used to describe and investigate their properties. The examples shown above are of less complex shapes. Note that the triple torus has three holes, as does the needed 6-d Calabi-Yau shape.

Since no one has ever observed any such dimensions, either directly with our senses or indirectly with our latest experimental probes, they would have to be very small indeed. Using the ideas and mathematics of quantum mechanics, Klein calculated the size of the circular dimension to be, amazingly, about the Planck length. Soon, it was demonstrated that there was no reason to limit the number of extra dimensions to one. Although the one-dimensional circular dimensions did not work out, their trial did indicate that added dimensions might have a place in string theory. See Figure 2.13.

As you will remember, a lot was happening in physics during the twenties. The focus and progress was on quantum theory. This dominance of quantum mechanics was maintained by its amazing agreement with data from trusted experiments. As a result, those physicists working with extra-dimensional theory were hardly noticed.

Eventually, however, the two found each other. The discoveries in quantum mechanics met up with extra dimensions. Finally, It was realized that the Kaluza-Klein attempts were probably not broad enough to handle the amount of information developed by quantum research.

Beginning in the 1970's, different numbers of curled-up dimensions were tried. The most successful also included supersymmetry. Some of these theories were called higher-dimensional supergravity theories. These theories seemed to be getting close. Some of the equations were similar to those of the strong and weak forces and the electromagnetic force. However, something seemed to be preventing the unification of the force and matter particles into one overall theory. The problem seemed to be related to the quantum jitter of space; the jitter was still too large. Since the above theories were based on the standard theory, could it be that replacing the point particles with loops of string would resolve the problem?

Extra Dimensions and String Theory

As we discussed earlier, string theory provides a solution to the incompatibility of general relativity and quantum theory. Astonishingly, calculations in string theory appear to indicate that *seven* extra spatial dimensions are *required*. At first, the approximate calculations seemed to indicate six extra spatial dimensions, but later seven was determined to be correct, although one of the seven may be different. It may provide an extra dimension to strings (thickness) or another large special extended dimension.

The number of dimensions of string theory arose out of calculations that yielded negative probabilities except when strings were vibrating in ten spatial dimensions. That would be a total of eleven dimensions…one time, and ten spatial dimensions. Since the determination of the need for seven extra dimensions didn't occur until about 1995, superstring theories continued to use six; and you will find that all but recent sources reflect the earlier number. In addition, M theory may require an additional extended dimension, in which case there would be only six curled-up dimensions, four extended dimensions and time. At this time, this question has not been settled.

Resonance

You will recall that the vibration resonance pattern (including spin) of a string determines both the mass and charge of the particle it produces. Just as the resonance chambers of musical instruments influence the resonance of music; the shape of the curled-up dimensions in which a string is vibrating influences the resonance of the string. As an example, you can think of strings as vibrating inside resonance chambers within the curled-up dimensions. This leads us to believe that the specific shape of the curled-up dimensions is crucial to string theory.

THE SHAPE OF THE CURLED-UP DIMENSIONS

We have seen that the resonance patterns of the vibration and spin of strings determine the masses and charges of the twelve matter particles (and their anti-matter partners). In addition, the force particles are determined in the same manner. See Tables 1 and 2. You will recall that the matter particles fall into three families. The four matter particles in each family differ in mass, electric charge and a couple of other properties.

All matter particles discovered to date are made up of some combination of the twelve particles in these three families. That includes a number of particles that have been found only in the detectors of colliders. It also includes other particles such as neutrinos, which are constantly passing through everything around us, including our bodies. This, and the fact that some of the twelve particles exist only under extreme conditions, helps explain why there are so many particles. Recall, that only the electron, up-quark and down-quark are needed to form the ordinary matter particles of our environment. Moreover, only one fundamental component, the vibrating loop of string, makes up both quarks and electrons.

We would expect that fine-tuning would be needed to produce the exact frequencies required (after subtracting for quantum jitter) to produce the masses and charges of the fundamental particles. We would also expect that, as with musical instruments, the design of the resonance chamber would be critical to the fine-tuning. In addition, since the resonance chamber for a vibrating string is within the curled-up dimensions, the precise design of the curled-up dimensions is critical to the whole process.

Calabi-Yau Shapes

In 1984, string physicists demonstrated that a certain class of six-dimensional geometric shapes meets the basic requirements for the shape of the curled-up dimensions. (This was more than a decade before it was discovered that the number of dimensions required by string theory is probably seven, rather than six.) These six-dimensional shapes are known as *Calabi-Yau shapes* and are used in a number of string theories.

The mathematics describing Calabi-Yau shapes is tedious and difficult; but it is not essential to our level of understanding of string theories. One discouraging fact, however, about this class of shapes is that there are tens of thousands of different Calabi-Yau shapes that seem to fit the general requirements of string theory. Obviously, one way to reduce the number would be to develop more specific requirements for the one shape we are seeking.

Resonance Chambers

One such additional requirement is the number of holes in the shape. A typical Calabi-Yau shape has holes similar to the holes in the triple donut (torus) shape illustrated in Figure 2.13. Theoretical physicists studying the effects of such holes on strings, vibrating within Calabi-Yau shapes, noticed something very interesting. They found that those with *three holes* should produce *three families* of particles.

Needless to say, the requirement of three holes was added to the specifications for a shape. [Note that this is a case of experimental data (three families) being used to aid the development of a theory.]

Thus, the number of candidates for the *one* needed Calabi-Yau shape was reduced significantly. After many tedious calculations, several Calabi-Yau shapes with three holes were found that theoretically would produce resonant vibrations close to those needed. Nevertheless, there are still no clear mathematical equations for determining the *one* precise shape; and it is believed that shape has not been determined, as yet.

You are probably wondering why we are dealing with six-dimensional Cabli-Yau shapes since we now believe that superstring theory probably requires seven-dimensional curled-up shapes. It is because the problem of determining the needed shape should be similar with seven-dimensional shapes. In addition, it is possible that six-dimensional shapes may still apply; and the seventh extra dimension is special in some way. There is little information currently available on research into seven-dimensional shapes.

Such research did begin in earnest with the introduction of M theory, which also requires curled-up dimensions. This research has uncovered a category of seven-dimensional shapes that seem promising. They are called Joyce manifolds. That is the shape that *may* apply in Section 1 for the "particles of space" if a Calabi-Yau shape doesn't turn out to be the one. Recall that space particles were assumed to fill the three-dimensional volume of the extended space of our visible universe. In any case, the space particles are assumed to be the same as the shapes that can house the vibrating strings that produce the matter and force particles of string theories.

THE SUPERSTRING THEORIES

Prior to the second superstring revolution, there were five viable versions of string theory, named as follows: Type 1, Type 2 A, Type 2 B, Heterotic O, and Heterotic E. All five share many key features such as ten or eleven space-time dimensions, including six or seven curled-up dimensions. In addition, they all have vibration patterns that determine mass and force charge.

Although these five theories have a lot in common, they do seem to differ. These differences relate primarily to the value of the coupling constant (will be defined later) of each theory, as well as how they include supersymmetry.

Duality

A common problem with the five theories is that each has many solutions. Moreover, most of the solutions do not seem to relate to our universe. Concentration on this problem during the second superstring revolution revealed that the "many solutions" problem was related to the approximate nature of the equations being used. The complex mathematics is still a problem.

However, discovery of a property known as *duality* between theories, enabled theorists to devise a strategy to partially overcome the approximation problem. Duality means that although the five theories appear to be quite different, they are similar when certain parameters are varied and the unifying central area of M theory is included. See Figure 2.14. Physicists are confident that all six theories (including the 11-d supergravity theory) fit into the overall framework of M theory.

Unification Under M Theory

It appears that M theory has emerged as a unifying theory. This unification of the string theories means that we can go from any one of the six theories to any other, if we also include the unifying M theory. See Figure 2.14. Thus M theory serves as a master theory interrelating all of the viable string theories.

M theory also provided some surprises for superstring physicists. It allows membranes of zero through nine dimensions. We now have a theory that includes not only strings, but also "branes" of from zero to many dimensions. However, when we are in the areas of the five string theories, only strings, or membranes curled-up like strings, seem to be able to produce masses light enough to match the fundamental particles and forces that are so well documented by experimental results.

The additional dimension of eleven-dimensional M theory was missed in the five string theories because of the approximation techniques used in their calculation. As we will see soon, the eleventh dimension, the principle of duality, the coupling constant, the principle of BPS states and M theory all led to a surprising new understanding of all of the string theories and their striking relationship to each other.

The Coupling Constant

One of the most important factors in the equations governing vibrating strings is a constant known as the *coupling constant*. This number determines how strongly the string loop is coupled (bound) to the two virtual loops into which it can split, producing quantum jitter. A larger coupling constant means there is a greater probability that the string will split into two virtual strings and then quickly recombine into the original string. Likewise, a smaller coupling constant means there is less probability that this process will occur. The quantity *one* is the dividing line between weak (less than one) and strong (greater than one) coupling constants.

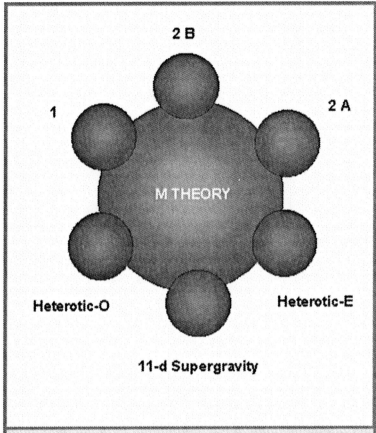

Figure 2.14 Unification Under M Theory. This diagram shows the relationship between M theory, the five string theories and 11-d supergravity. The dualities of the six theories indicate that each theory is a different interpretation of a part of M theory.

Recall that quantum jitter plays a major role in determining the effective mass of a vibrating string. The energy of jitter is *negative*, thus subtracting its energy from the *positive* energy of the resonant pattern of the string. In this way, jitter helps to produce the lighter masses needed for the fundamental particles and forces. Therefore, any string theory is highly dependent on its coupling constant, as well as the precise geometry of its six-dimensional curled-up shape (or seven-dimensional shape).

Despite their determined efforts, string physicists have not been able to directly compute the value of this constant for any theory. They do know, however, that approximation techniques work reasonably well when the coupling constant is less than one. For this reason, string physicists pretty much ignored coupling constants greater than one. That is, until Edward Witten lit a fire under them at that convention at Santa Barbara, in 1995.

BPS States

The organizational structure of supersymmetric theories can give clues to the identity of some particles that should be produced by vibrating strings. This gives at least some information about theories with strong coupling constants. A vibrating string that should produce a minimum mass and a specific charge identifies the particle that would be produced.

This principle was discovered by physicists Bogomoln'yi, Prasad and Sommerfeld, and is known by their initials as the *BPS state*. The BPS principle holds true for all values of the coupling constant, even those greater than one. Thus, it enables theorists to determine which particles with a minimum mass and specific charge should be produced. This gives us at least some information about theories with strong coupling constants.

The Duality of String Theories

In 1995, Witten proposed that the principle of duality could apply to the half dozen string theories. Recall that duality means that two or more theories that are somewhat similar in their construction and properties may actually be *identical in their results and yield the same physics*. It was suspected that the coupling constants might be at the root of the apparent differences in the similar theories.

The suspicion proved to be correct. Before long, it was demonstrated that the Type 1 string theory, with a strong (greater than one) coupling constant produces exactly the same masses and charges as the Heterotic O theory, with a weak coupling constant. Thus, these two theories were shown to be *dual theories*. Moreover, this means that the results of calculations that are easily approximated in one theory can be applied to its dual theory, where calculations are extremely difficult or impossible.

Although most string physicists are confident that this duality relationship is true, at this time it has not been proven. The equations are too rough, and the math is not specific enough for rigorous proof. However, using the principle of duality, it was demonstrated eventually that all of the five theories, along with M theory, are dual to each other. In addition, it is believed that a sixth theory, eleven-dimensional supergravity, is an approximation, for low energies, for the broader M theory.

THE STATUS OF M THEORY

By bringing the five string theories and eleven-dimensional supergravity under the broad purview of M theory, we have concisely summarized the current string theories. See Figure 2.14. However, at this time, we don't know a lot about M theory itself. We do know that M theory includes more objects than the one-dimensional strings assumed in the string theories.

It appears that these objects can range from zero-dimensional to nine-dimensional membranes. In the terminology of M theory they are called *branes*. The one-dimensional string is a one-brane. The membranes are two-branes, three-branes and so on. The general term is p-brane, where p is the number of dimensions.

Since the masses produced by the branes, other than the one-brane, are inversely proportional to the coupling constant, the higher-dimensional branes would produce very massive particles when the coupling constant is weak, that is less than one (dividing by a constant less than one yields a larger number). Therefore, the amount of energy needed to produce particles with such vibrating branes would be enormous, and would not produce any of the light fundamental particles.

Remember, however, that this is true only for *weak* coupling constants. It appears that at least some of the higher dimensional branes of M theory (in the center portion of Figure 2.14) should have *strong* coupling constants. If so, the particle masses would be much lighter (dividing by a number greater than one yields a smaller number). Thus, it is possible that branes of up to several dimensions in M theory are important to full understanding of the power of the theory.

The potential richness of M theory is sure to surprise us as it is rigorously explored and tested. We can look forward to some exciting discoveries and a broad fundamental theory of our universe. Perhaps, it will be possible to experimentally confirm predictions of the theory sooner than is expected at this time. Some of us believe it is about time for an Einsteinian breakthrough. We will continue to refer to the theory that is evolving from the merging of aspects of the superstring theories with M theory as the superstring-M theory (SS-M theory).

Section 3

The Proof and Potential of SS-M Theory

9

Building Credibility

As superstring-M theory continues to explain otherwise unexplained phenomena, the credibility of the theory increases. In the following discussions we will see how the theory adds to our understanding of a long-standing puzzle of astrophysics…black holes. In addition, we will see that the fundamental structure of black holes is the same, regardless of size, from astronomical to ultramicroscopic black holes. This is but one of a number of illustrations of the power and growing credibility of superstring-M theory.

BLACK HOLES

If a blob of matter is crushed by its own gravity, (without igniting nuclear fusion) into a smaller and smaller sphere, a space-time warp occurs. (You may want to re-read the black holes material in Section 1 and Figure 1.20, at this time.) The spacetime warp (gravity) can be so extreme that the sphere is transformed into a black hole.

Anything that comes too close to the black hole's event horizon will be consumed. Evidence of matter falling into a black hole is illustrated in Figure 3.1. The matter of a black hole is crushed to a density, difficult to imagine. For example, if the matter of our sun were compressed from its diameter of 900,000 miles to a diameter of only 4 miles, it would become a black hole. Such densities seem impossible. Remember, however, that matter is mostly "empty space". There is a lot of space between molecules, between the atoms that make up the molecules and between electrons and the nuclei of atoms.

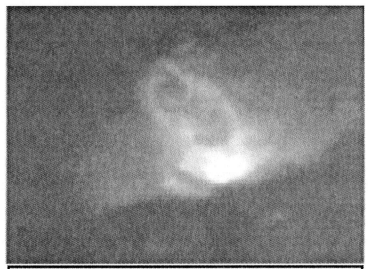

Figure 3.1 Evidence for Black Holes. This photo shows a black hole blowing monstrous bubbles of gases into space. The bright spot is an emerging bubble. It is about 800 light years in diameter. Above the bubble is a "smoke ring" from an earlier bubble. The glow is caused by cooler gasses colliding with a wall of dense hot gases. The bubbles are caused by the black hole's voracious consumption of matter from a nearby disc of gases and dust. Some blobs (bubbles) spin away from the event horizon and escape the pull of the black hole.

From One Extreme to the Other

In the late 1960's, a group of theoretical physicists determined that the structure of black holes is amazingly simple. They had noticed that all black holes seemed to be very much alike. Moreover, it appeared that, except for three properties, there isn't much to the basic structure of black holes. Then they realized that black holes that have the same amount of each of the three properties, *force charge*, *mass* and *spin* are identical. In addition, those that differ in one or more of the three properties are different; but they are still quite similar.

If all of this sounds familiar it is because you have heard it before. Yes, the par-ticles of string theory also have force charges, mass and spin. Could it be that large black holes are gigantic elementary particles, and that elementary particles are ultramicroscopic black holes? Well, relativity puts no minimum limit on the mass of a black hole.

Realizing this, theoretical physicists decided to see if they could create, *theoret-ically* of course, small black holes. Using the rules of relativity and spacetime geometry, they "crushed" smaller and smaller bits of matter to the required den-sity and "produced" black holes. Then they compared the properties of the tiny black holes with the properties of elementary particles. They found that, indeed, both have the same properties: force charge, mass and spin. However, a serious problem arose when they tried to crush bits of matter smaller than a certain mass.

When the bits of matter used in calculating a black hole reached a certain mass (the smallest they had tried) the formulas would no longer yield sensible values. Guess what that mass was? Right, the formulas wouldn't work at, or less than, the *Planck mass*. Since the Planck mass is billions of times larger than even the larger particles, such as the proton, this was a major problem. However, it was soon realized that the problem was related to the incompatibility of general relativity and quantum theory. After all, the threshold had been crossed into the domain of quantum mechanics; but the calculations also involved relativity.

I'm sure you know what the problem was. The formulas involved both quan-tum mechanics and relativity; and they don't mix, except in superstring theories. Since the theoretical calculations were made using the standard theory, they didn't yield sensible results. Progress was stalled on resolution of the black hole particle question until the mid 1990's.

Avoiding Catastrophe

For some time, another problem had been haunting string physicists. It seems that the approximate equations were indicating that a three-dimensional sphere embedded in a Calabi-Yau shape could collapse as the result of a random energy fluctuation. Furthermore, the equations indicated that the collapse might bring about a cataclysmic end to our universe. Not a cheerful thought.

However, you will recall that M theory entered the picture in the mid 1990's. As string physicists studied M theory, they soon realized that the membranes allowed in M theory could broaden the one-brane string theories. Then physicist Andrew Strominger demonstrated that a three-brane could wrap completely around a three-dimensional sphere. Moreover, the three-brane would shield the pinched spacetime from rupture. This is a little like patching a hole in a bicycle tire, just before the puncture occurs.

Ultramicroscopic Black Holes

Well, guess what (again). It was soon realized that a three-brane wrapped around a three-dimensional sphere would look like...*a black hole!* It was also realized that as the wrapped sphere collapses, the mass of the wrapping membrane (and black hole) decreases. Moreover, as the sphere collapses to a point, the black hole becomes massless! Well, that's a new critter...a massless black hole? What in the world is a massless black hole?

Through some clever and involved work on this question along with more study of space-tearing conifold transitions, by physicists Greene, Horowitz, Morrison and Strominger, a resolution to the problem was found. It was established that when such space tearing occurs, the number of massless string vibrations increases by one. Since this same kind of space tearing occurs as a black hole shrinks to a point, it was realized that as the black hole approaches point size it transforms into a massless vibrating string! The string can be a photon or other massless particle. Paradox solved.

10

The Search for Proof

Relating astronomical black holes to black holes at the elementary particle level demonstrates that superstring-M theory is a theory for all sizes. This, and numerous other explanations of unexplained phenomena that superstring theory can resolve, has built credibility. However, credibility is not proof. Next, we will look at several possibilities for finding experimental proof.

LOOKING FOR SUPERPARTNERS

Because of the theoretical and mathematical problems related to superstring theory, it is not possible to make specific detailed predictions from the theory at this time. However, there are some properties of the theory, which make a few predictions possible because of the contributions the properties make to the theory.

As you know, the property of supersymmetry makes major contributions to superstring theory. It requires, for example, that when matter particles are produced they must be produced in pairs. So, for every fundamental particle there must be (or have been) a partner...a superpartner. Furthermore, the partners must differ from each other by ½ unit of spin. Therefore, the theory predicts that there must be superpartners.

As we discussed earlier in the topic *Superpartners*, no such particles have been found. It is believed that they are too heavy to be produced in current colliders. However, several new or up-graded colliders are being readied for the search. Discovery of superpartners would be a dramatic event in particle physics. You can join the excitement by visiting collider laboratory websites. Fermilab, in Chicago, could be the first to observe a superpartner, if any are light enough. CERN in Geneva, Switzerland, is preparing its large proton collider to join the search soon.

It should be noted that in these experiments enormous quantities of data are collected over a period of months or more. In addition, quite a long time is required to analyze the data. Remember too, that finding superpartners would not be complete proof of superstring theory. However, it would confirm that supersymmetry is a part of nature; and that, therefore, any successful theory of matter will have to be supersymmetric.

PROTON DECAY?

It had been believed until recently that protons are stable and do not decay. In fact, the standard theory doesn't allow proton decay. With the addition of supersymmetry to theories, including superstring theory; it was realized that protons might decay. If such decay should occur, it would happen very slowly. For example, the half-life of protons might be as long as the lifetime of our universe. (The half-life of particles that decay means that, in any sample of the particles, half will decay during the half-life period. Half of the remaining half of the particles will decay during the next half-life period, and so forth. See Figure 3.2. Experimentally observing proton decay seems hopeless until we realize that there are a colossal number of protons in our universe. They are present in the nuclei of all atoms. This means that if we could observe a large number of atoms for a long enough time, it might be possible to detect a decaying proton, now and then.

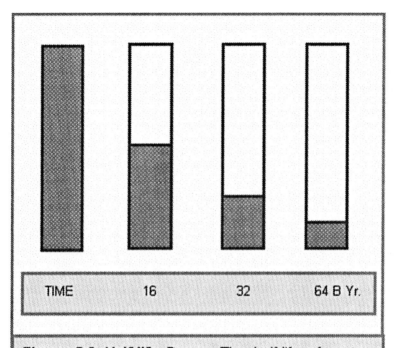

Figure 3.2 Half-life Decay. The half-life of proton decay has to be an extremely long time. One half of any sample of protons decays during the half-life period. In the example above, a half-life of 16 billion years is *assumed.* Each 16 billion years half of the protons decay into something else, if protons do decay.

The challenge was accepted by a number of experimental physicists. The first proton decay experiment was set up in a salt mine near Cleveland, Ohio. The experimental equipment is amazing. Extremely pure water provides the protons. Since the probability of decay is so small, a huge amount of pure water is necessary. The Cleveland tank is about the size of an Olympic swimming pool. The sides of the water tank are lined with photomultiplier tubes designed to catch any photons emitted from any proton decay event that may occur.

When all was ready, the waiting game began. Because of the odds, the researchers would have been happy to record one or two decay events in a year. Therefore, it is not surprising that at the Cleveland lab no proton decays have been observed, so far.

However, there may be another reason. After the Cleveland experiment had been running for some time, theorists realized that they should have included supersymmetry in the theories used in calculating the half-life of proton decay. New calculations indicated that the half-life is a thousand times larger. This means that the experiments must use more water and time, if proton decay events are to be detected.

You would think the physicists would give up. However, physicists are a determined and patient lot. A much larger tank was set up in Japan, and it is being watched at this time. The size is believed to be sufficient to cover enough of the half-life so that there is a fair chance it may record a decay event. Meanwhile, in Italy, another proton decay detector is being planned.

Someday, proton decay may be detected. If so, it will prove supersymmetry in nature and strengthen SS-M theory. However, since we don't know the exact formulas of SS-M theory, the failure to find proton decay would not disprove the theory. If, and when, we do know the formulas, they predict proton decay, and we still don't find it; then there will be a problem with SS-M theory. If the failed experiments were not definitive, new attempts to find proton decay would be in order.

THE ELUSIVE NEUTRINO

Neutrinos are blasted out of the fusion furnaces of stars at almost the speed of light. They travel in straight lines right through matter, as if it weren't there. Recent experiments seem to indicate that neutrinos do have a tiny size and mass. However, both the size and mass are *extremely small*. For example, if a neutrino were magnified to the size of a grape; an atom magnified the same amount would be about ten times the size of the earth. Neutrinos carry the weak charge but no electric charge. Thus, they cannot be deflected or controlled by electromagnetic fields.

This makes it difficult to measure a neutrino's mass or control it's path, as can be done with charged particles. They can interact with other particles, but this happens only if the neutrino gets *very close* to another particle. This is because the range of their weak force is limited to nuclear distances.

The Probability of Neutrino Interaction

The weak force is the mechanism of radioactive decay; and neutrinos are produced by decay processes, such as beta decay. For example, beta decay occurs when a neutron emits a W boson and transmutes into a proton, an electron and a neutrino. Although a neutrino can interact with another particle and transmute into something else, this *almost* never happens. This is because *matter is mostly empty space*. Moreover, since most of the neutrinos do not come close enough to be attracted to any other particle, they just shoot on by at near light speed.

However, if a neutrino and another particle do come within range of their weak forces, they *may* interact. The probability of this occurring is *almost zero, but not quite*. This small chance makes it possible to detect neutrinos indirectly, by observing any electrically charged particle emitted in the interaction process.

Another reason why neutrinos almost never interact with other particles is because W and Z bosons are very massive (heavy). They are about a *hundred billion* times as massive as neutrinos. W and Z bosons are the force particles that transmit the weak force. This means a neutrino has to *emit* a boson to interact.

Well, that must be about as impossible as anything could be. Could you pass a basketball, which weighs a hundred billion times as much as you do, to a teammate? The answer is obvious. Remember, however, that the neutrino and boson are in the quantum realm where Heisenberg uncertainty allows a *low probability* that it can happen. *Given enough neutrinos and enough time*, the event should be detectable.

Some History of the Neutrino

Now that we know some properties of neutrinos, it is not surprising that they are so elusive. Wolfgang Pauli postulated the neutrino, in 1930. Pauli was puzzled by the missing matter-energy in the beta decay of the neutron (not neutrino). Measures of the matter-energy before and after the process should be equal. Rather than dispute this well-established law of the conservation of matter-energy, Pauli postulated that an *undetectable particle* is emitted in the decay process.

A few years later, Enrico Fermi developed a mathematical description of beta decay and named the new particle the *neutrino*, which means "the little neutral one" in Italian. The neutrino was not experimentally confirmed until 1958. Now we know that one fourth (three types) of the twelve elementary particles are neutrinos. See Table 1. If they have any mass at all they might outweigh all the rest of the ordinary matter in our universe! It is believed that neutrinos may account for at least twenty percent of the missing matter-energy. Of course, this estimate depends on the number of neutrinos in our universe, and that is only an educated guess at this time.

The Origin of Neutrinos

Neutrinos were produced in the aftermath the big bang, and many of them are still shooting through the universe. In addition, the fierce fusion reactions in the cores of our sun, the other stars of our galaxy and the other galaxies produce copious quantities of neutrinos. The *number of neutrinos is so colossal* that it is truly indescribable. They are constantly streaming through space, the earth, our homes and our bodies.

They don't appear to harm us. Moreover, that's a good thing, because about a thousand trillion neutrinos shoot through our brains, unnoticed, each second. Some of them pass through the earth first. Others pass through us first and then on to the edge of the universe. Neutrinos are just about unstoppable.

Some Neutrino Research

Nevertheless, that hasn't stopped more than a thousand physicists, from all over the world, from trying. Several huge detectors are being built, usually with the combined funds of several nations. There are two basic types of experimental facilities. One type is simply a neutrino detection device that is designed to detect neutrinos in the stream from space. The other type produces a stream of neutrinos in accelerator-collider equipment. This allows the production of one kind of neutrino only, such as the muon neutrino. The neutrino stream is fed to a suitable detector. All detectors must be far enough underground to shield them from cosmic rays, which would contaminate the data.

Neutrino Detectors

The first type of detector consists of a huge tank of pure water lined with photosensors to record any interaction of a neutrino with the water. Neutrinos can be detected, only indirectly, by their interaction with a particle of matter. An interaction usually produces a particle, such as photon or electron, which can be detected. As we saw earlier, these interactions have a very low probability. Therefore, even if the experiment goes well, it is expected that only a few events will be recorded in a year of operation. Facilities with this type of equipment are operating in Sudbury, Ontario and Kamioka, Japan. The other type of equipment is at CERN in Switzerland and Gran Sasso in Italy. An accelerator-collider is being prepared for producing a stream of *muon* neutrinos at CERN. Meanwhile, two detectors, of different types, are being built at Gran Sasso, Italy.

The plan is to produce a stream of *muon* neutrinos at CERN and shoot them in a straight-line underground, through the rock and earth, to detectors in Gran Sasso, 454 miles away. The experiment is designed to test the theory that neutrinos can change from one kind to another as they speed on their way to wherever.

Some Properties of Neutrinos

As you know, there are three kinds of neutrinos. See Table 1. They are the *electron* neutrino, *muon* neutrino and *tau* neutrino. They are named for the particle emitted when on rare occasions they interact with another particle. Currently, at Sudbury and Kamioka, experimental data indicates that neutrinos do have a tiny mass. In addition, the data suggests that a specific neutrino does not maintain a constant mass value or identity.

Rather, as it shoots through space, it *oscillates* from one identity (kind of neutrino) to another. As physicists would say, the neutrino can oscillate from one *flavor* to another. The flavor of a neutrino indicates the kind of particle that will be emitted, if it gets close enough to a particle for an interaction to occur. As shown in Table I, there are three flavors of neutrinos.

In the late 1930's, an Italian physicist, Ettore Majorana, hypothesized that the neutrino is its own antiparticle. If this is true there should be a double beta decay in which the antineutrino emitted by one neutron is absorbed by a second neutron. Then the two neutrons would decay without emitting any neutrinos. If this is found to be true, it will have a major effect on particle theories. Currently, experiments are being set up or are underway in France, Spain and in a Heidelberg-Moscow collaboration. Now, some six decades after Majorana proposed the self-antiparticle hypothesis for neutrinos, the world's physicists are scrambling to test his idea; and they are wondering also if it might apply to other matter particles.

The Cern-Gran Sasso Experiments

The CERN-Gran Sasso experimental set-up is remarkable. See figure 3.3. Muon neutrinos are produced by shooting protons, traveling at ninety-nine percent of the speed of light, into a special graphite target. The target is designed to aim the muon neutrino beam that is produced precisely enough to inject the beam into a detector at Gran Sasso, 454 miles away. (A tunnel is not needed.) There, two different types of detectors await any *tau* neutrinos that may have oscillated from muon neutrinos during the trip.

One of the detectors is ICARUS. It consists of four huge, rectangular, double-walled aluminum tanks. On the inside walls of the tanks are three layers of thin wires…twenty-six thousand in all. The tanks are filled with liquid argon. Argon is used because it is an inert element and does not interact with other atoms or charged particles. Liquid nitrogen circulates through the double walls of the tanks to keep the argon liquid at—300 degrees F. If, and when, a tau neutrino hits an argon atom and interacts to produce a charged particle, the particle will be pulled to the wires by an electromagnetic field created to do just that.

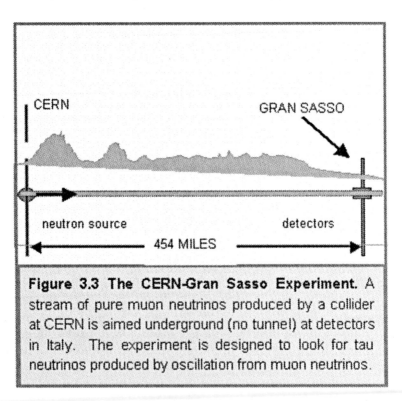

Figure 3.3 The CERN-Gran Sasso Experiment. A stream of pure muon neutrinos produced by a collider at CERN is aimed underground (no tunnel) at detectors in Italy. The experiment is designed to look for tau neutrinos produced by oscillation from muon neutrinos.

Remember, we are looking for evidence that a *muon neutrino has changed into a tau* neutrino. This detector is designed to detect evidence of tau neutrinos. The current from the thin wires is fed into computers that record and display the information.

The second Gran Sasso detector, OPERA, is more conventional in design. See Figure 3.4. It uses 235,000 lead brick sandwiches. Each brick is made of 58 rectangles of photo film sandwiched between rectangular sheets of lead. The 235,000 bricks are stacked to make 72 walls. In the space between the walls a plastic scintillator material lights up when a charged particle passes through a brick. An automatic mechanism identifies the brick and removes it for processing. When the film is developed, it shows the track of the particle that passed through the film. A computer analyzes the track and displays it.

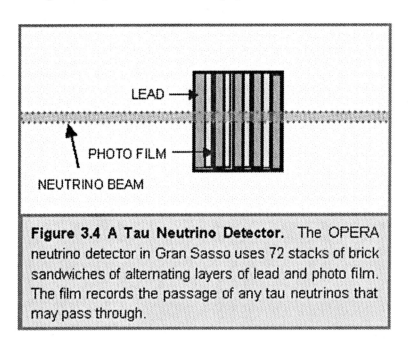

Figure 3.4 A Tau Neutrino Detector. The OPERA neutrino detector in Gran Sasso uses 72 stacks of brick sandwiches of alternating layers of lead and photo film. The film records the passage of any tau neutrinos that may pass through.

It is calculated that OPERA may detect evidence of a half-dozen tau neutrinos in a year of continuous operation. That would mean that a half-dozen, out of the million trillion muon neutrinos that passed through the walls, oscillated into tau neutrinos during the 454-mile trip from CERN. That would be definitive proof of the flavor oscillation of neutrinos.

POSSIBLE UNUSUAL PARTICLES

The twelve known elementary particles have electric charges of 0, 1, -1, 1/3, -1/3, 2/3, or -2/3. All electrically charged matter that we know about is composed of various combinations of particles with these charges. However, superstring theory *may* predict particles of other fractional electric charges.

These charges would be related to winding strings wound around holes in a specific Calabi-Yau shape. Only unwinding the strings a specific number of times can unwind them from these shapes. The number of times may be 5, 11, 13, etc. Calabi-Yau shapes, of this design, allow string vibrational patterns with fractional charges determined by the reciprocal of the unwinding number. In the example above, the fractional charges produced would be: 1/5, 1/11, 1/13, etc.

Like the superpartners, the masses of these particles cannot be computed; but they also are expected to be heavier than our current accelerator-colliders can produce. Again, several new or rebuilt accelerator-colliders may be able to produce and detect these particles if they do exist.

NEW FORCE FIELDS?

There are also Calabi-Yau shapes that allow the production of tiny long-range force fields. As yet, such fields have not been observed. However, new force fields would not be observed unless something reacts to the field; and the reaction can be observed.

We humans can easily sense some forces because we have detectors in our bodies that react to the force. For example, we can feel gravity. We may not want to, but we can feel electricity too, in certain situations. In addition, we can see light. However, we cannot detect magnetism without something that interacts with the magnetic field.

Once we knew that iron, and anything made of iron, interacts with magnetic fields we had a detector. Since iron is fairly plentiful on our planet, magnetism was discovered early in our history. However, if iron had been rare, the discovery of magnetism probably would have occurred much later; and the progress of physics and civilization would have been delayed.

The point is: *currently unknown force fields could exist.* The weak and strong forces were not known until recently. If a new, tiny, long-range force should be discovered, it would be strong support for SS-M theory. Could it be a repulsive force that would explain the increasing speed of expansion of our universe?

THE COSMOLOGICAL CONSTANT

As we discussed earlier, in the topic *Expansion of Our Universe,* in Section 1., during the last few years astrophysicists have found evidence that the expansion of our universe is *faster* now than it was several billion years ago. Since this discovery, cosmologists and particle physicists have been scrambling to find an explanation. One possibility is that the cosmological constant, lambda, is not zero. The zero value has been assumed since the big bang theory was generally accepted. At that time, a gradually *decreasing* speed of expansion was a part of the big bang theory. Currently, the search is on for an explanation of the cause of the *increasing* rate of expansion. We will examine a few possibilities.

Is Lambda Really Zero?

The above evidence has led to new investigations into the value of lambda. It seems there is no specific reason why lambda should be zero, except that it appeared to work for a gradually slowing expansion. Lambda can be thought of as representing energy stored in the vacuum of space.

Using this premise, calculations of the energy of quantum fluctuations in the vacuum of space were made. Experimental measurements were made also. Well, that didn't help much. The experimental measurements of lambda turned out to be zero, or just a bit more than zero. However, the calculations of the energy of quantum fluctuations yielded a value 120 orders of magnitude greater than the "bit more than zero" amount. That's quite a mismatch.

Knowing that the theoretical calculations were based on the standard theory, string physicists have suggested that use of the superstring-M theory might yield a more reasonable value for lambda. It is believed that some work is being done along these lines. In any case, the value of lambda needs to be settled. If string theorists can resolve this problem, it will add additional support to Superstring-M theory. As we will see near the end of this book, Alan Guth may have an explanation for the increasing speed of expansion built into his recent fluctuation/ inflation theory.

DARK MATTER

In 1938, Fritz Zwicky discovered an anomaly in a cluster of galaxies known as the Coma Cluster. He calculated that the cluster was spinning so fast that it should have come apart long ago. The amount of observable matter in the cluster was simply not enough to hold the cluster together. Not enough matter, meant not enough gravity to hold the rapidly rotating cluster together.

However, the cluster was not flying apart. Therefore, Zwicky hypothesized that *some unseen form of matter* was supplying additional gravitational force. That was seven decades ago, and the mysterious *dark matter* has not yet been identified. See Figure 3.5. However, the case for dark matter is rapidly becoming stronger. Furthermore, observations of a number of other rotating galactic clusters indicate that Zwicky was right.

Something unseen is helping to hold the galactic clusters together. In 2001, the Chandra X-ray Observatory made indirect, but very precise, measurements of the amount of dark matter that enshrouds a galactic cluster, four billion light years away. Those measurements indicate that *dark matter* makes up 80 percent of the total amount of matter in the cluster. This means that ordinary matter makes up only twenty percent of the mass (gravitational force) that holds the spinning cluster together. Thus, it appears that dark matter can explain how the cluster is held together. It also means that the case for the existence of dark matter is becoming quite strong.

Estimating the Amount of Ordinary Matter

More recently, astrophysicists have been able to measure the relative amount of deuterium (a hydrogen atom with a neutron in the nucleus along with the proton) in clouds of hydrogen gas. These gases of hydrogen and deuterium were formed soon after the big bang. The ratio of deuterium to hydrogen enables cosmologists to calculate the abundance of protons and neutrons in the universe.

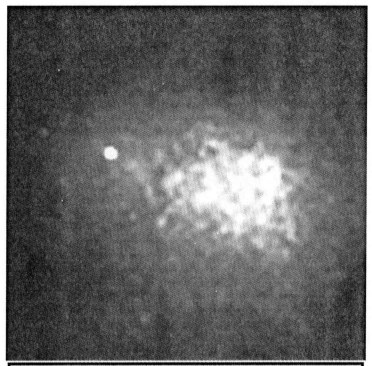

Figure 3.5 The Coma Cluster. The above picture is a Chandra X-ray view of the central region of the galactic Coma Cluster. This region is about 1.5 million light years across and contains thousands of galaxies. The galaxies are shrouded in a gigantic glowing 100 million degree cloud of gas. Embedded in the hot cloud are cooler gaseous clumps. It is believed that magnetic fields keep the hotter and cooler gases separated. The brighter spots are large galaxies that shine through the gaseous clouds.

Since the nuclei of all "ordinary" matter particles are made of protons and neutrons, the amount of ordinary matter in the universe can be computed. You will recall, from Section 1, that for omega to have a value of *one* the universe must contain a specific amount (density) of matter-energy. You will recall also that a number of experiments mapping the patterns of density of the microwave radiation from the big bang indicate that omega is one. However, the calculated amount of ordinary matter in the universe appears to be less than 4 percent of the omega one amount.

If all of the above is true, it means that ninety-six percent of the matter-energy of the universe is unobservable by the usual means. It also means that determination of the properties, structure and amount of dark matter is vital to both cosmology and particle physics.

Kinds of Dark Matter

Several types of dark matter have been proposed. One, *cold dark matter*, was the first model and has had some success. The properties of this form of dark matter are inferred, partially, from the properties of ordinary matter that dark matter doesn't have. Dark matter doesn't radiate observable light, it doesn't move fast and the particles are not light compared to ordinary matter particles. Dark matter exhibits only two forces: the weak force, which is limited to nuclear distances, and the gravitational force. This description of dark matter fits pretty well with what was *not* observed and what *was* observed in early observations.

The cold dark model may explain how some galaxies formed so soon after the big bang. It is believed that, at that time, the sluggish, heavy dark matter particles would have gathered into blobs earlier than energetic protons and neutrons would have. Thus, the shape and spatial arrangement of these early blobs of dark matter could have provided a 3-D matrix (framework) on which ordinary matter formed. Ordinary matter would have been drawn to the dark matter by its strong gravitational attraction.

Recent observations seem to contradict some predictions of the cold dark matter model. One such observation indicates that the centers of the galaxies are not as dense as the model predicts. In response to this observation, a *warm dark model* was developed. In this model, the warm dark matter particles have enough energy to bump into each other faster and more often. This would offset enough of the gravitational pull to reduce the compaction of the particles, and thus reduce the density of the center area of the galaxy. There are at least a couple of other dark matter models. However, the above examples are enough to give us a basic understanding of what dark matter is all about.

Surveys of some hundred thousand galaxies indicate that the pattern of distribution of dark matter does match the distribution of galaxies. Thus the role of dark matter in providing a matrix for the formation of galaxies, as well as additional gravity to hold rapidly spinning clusters together, seems to be pretty well established. The surveys also indicate that dark matter probably can account for about 23 percent of the matter-energy necessary for omega to be one. Ordinary matter can account for about 4 percent. Something else is needed to make up the balance of 73 percent.

SS-M Theory and Dark Matter

Now, let's see what SS-M theory has to offer on the problem of dark matter and the missing 73 percent of matter-energy. We don't really know for sure what happened to two categories of superstring particles. They are the massive particles and the superpartners. We have discussed possibilities and probabilities, but we don't have enough information about these particles to determine if they are dark matter particles.

If either of these particles are produced and detected in accelerators, we should be able to determine if either is a candidate for dark matter. In any case, there is a good possibility that there are string vibration patterns that could produce particles with only the gravitational and weak forces. Any of the above events would demonstrate the breadth and power of SS-M theory.

DARK ENERGY

As we learned earlier, *dark energy* also has been hypothesized to account for some missing matter-energy. In addition, dark energy has been considered as an explanation for the increasing rate of expansion of our universe. Dark energy would have to *repel* the objects of our universe enough to exceed the gravitational attraction at large distances. The amount of this dark repulsive force necessary to speed up expansion to the observed speed would be critical.

Although several proposals for dark energy have been made, they are highly speculative at this time. The most promising theorizes that dark energy could reside in the vacuum of space. However, as noted above, calculations based on the standard theory produced a force some 120 orders of magnitude greater than the critical force needed to match the observed expansion.

Calculations based on SS-M theory should yield a different result. At this time, astrophysicists are carefully checking the accuracy of their measurements of the expansion rate of our universe and looking for evidence of the existence of dark energy. *Quintessence* has been proposed as a name for dark energy. Even if we accept dark matter the still missing 73 percent of matter-energy is a major problem. Theoretical physicists are hard at work in search of a solution.

A STRING IN THE SKY?

The length of the strings that produce the vibrational patterns of the elementary particles of string theory is set at about the Planck length. However, not all strings are that size. As was noted in the topic, *Winding Strings*, the longer a string must be to wind around a hole in a Calabi—Yau shape, the more energy it has. In addition, not all open-ended strings are the same length. In fact, their length can *grow* if abundant energy is available.

Well, energy was *very* abundant at the time of, and for some time after, the big bang. Therefore, it has been suggested that some strings could have grown very large during that time. Moreover, some of the large strings could have escaped later break-up processes. Then, as the universe expanded, these strings could have grown to astronomical sizes. Therefore, there is a small probability that such huge strings still exist. If so, someday evidence of such strings may be found by our sky surveillance satellites or telescopes. Although this may be the least likely scenario for finding proof of superstring-M theory, it would certainly be the most dramatic.

11

The Potential of SS-M Theory

THE CONCEPTUAL POTENTIAL

The idea of a vibrating string (a one-brane) as the fundamental unit of matter and the forces is a powerful concept. It is simple but broad in application; and it can bridge the gap between general relativity and quantum mechanics. The replacement of the point particle with tiny vibrating strings is at the heart of the explanatory power of string theory. In addition, the inclusion of branes of other dimensions, as provided by the more recent M theory, is expanding the explanatory power of the evolving SS-M theory. Thus, as these theories are modified and merged, the power of SS-M theory should continue to increase. SS-M theory may well evolve into a "theory of everything" in physics.

Since the laws of physics are also the laws of our universe, they are the primary laws of the other major sciences, such as biology, geology, and physiology. Proof and a better understanding of SS-M theory would lead to a flood of new knowledge in all of the sciences. Let us hope that this new knowledge will be applied to the technology of our civilization in a manner that will benefit humankind.

A NEW ENERGY SOURCE?

The vibrating strings and branes of SS-M theory represent energy. The fundamental particles of ordinary matter, the up-quark, down-quark and electron, may be thought of as either energy or matter (mass). One of these, the electron, is readily available and easily manipulated with electric and magnetic forces. We do it all the time, as is illustrated by the multitude of electric and electronic devices so important in our daily lives.

These devices use electrons as "pack horses" to carry energy of motion. Electromagnetic forces push electrons through wires, a vacuum and printed circuits.

Electrons flow through various devices designed to produce light, heat, sound, images, information storage, heating, cooling and motion in machines of all kinds. However, in none of these cases are the electrons, themselves, the *source* of the energy. The energy that the electrons carry in electric current comes from burning fuels, chemical reactions in batteries, nuclear fission, falling water and the sun.

Since electrons are very plentiful and easy to manipulate, is it possible that their mass could be converted directly into energy? Electrons are energy...vibrating strings. The amount of energy in an electron is easy to determine. We just use the familiar $E=mc^2$. The energy is calculated by inserting the mass of the electron and the speed of light squared in the formula. We will not go through the full calculation because of the need for measurement terms with which the reader may not be familiar.

It is easy to see, however, that an extremely tiny amount of matter will yield a comparatively huge amount of energy. That is, of course, because the mass of matter *is multiplied by a colossal number.* That number is the speed of light *squared*: 30,000,000,000 centimeters per second squared. In other words, $(30 \times 10^9)^2 = 900 \times 10^{18} = 900,000,000,000,000,000,000.$ or, 900 quintillion times the mass unit. Calculations show that the energy equivalent of the proton is about 1,000 kilowatt-hours. Therefore, the energy equivalent of the electron is about .54 kilowatt-hours (.00054 X 1000 kwh=.54 kwh).

Thus, it is easy to see that if we could persuade electrons to give up their intrinsic energy, we would have a new and plentiful source of energy. This wouldn't be nuclear fission or fusion. It would be *annihilation.* See Figure 3.6. It can be done, and you probably know how. Yes, just squirt some electrons into a suitable chamber containing positrons; and the matter and anti-mater would annihilate each other in a burst of pure energy.

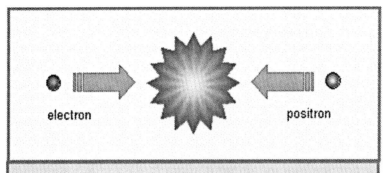

Figure 3.6 Annihilation. When an electron (-) and a positron (+) are near enough, they come together in a blast of energy. Since they destroy each other, the energy released is from the conversion of matter into energy in accordance Einstein's formula.

There are, of course, problems in achieving this process. However, it does occur in the vacuum of space and in accelerator-colliders. Obviously, one major problem is that unlike electrons, positrons are not readily available. Another problem is that the energy released by a virtual particle pair in the vacuum is almost instantly repaid to our universe. In addition, collider machines are extremely cost inefficient. As is the case with fusion energy, the problem might be one of cost efficiency.

On the other hand, it might not. It is quite possible that the equations and knowledge emerging from a mature version of SS-M theory could lead to currently unknown ways to utilize the energy stored in vibrating branes. Could it be only a matter of cutting the string loop that is the electron? Probably not. In any case, the primary purpose of the above speculation is to provide a hint of the potential for technology of a mature, proven SS-M theory.

BLACK HOLES AND INFORMATION STORAGE

In 1974, Steven Hawking developed a formula to determine the *entropy* of black holes. Entropy is the measure of the number of different ways the objects inside a black hole can be arranged without changing the outside appearance of the hole. The formula indicates that the entropy of a black hole is directly proportional to the *area of the black hole's event horizon.* This is very intriguing because it appears that somehow the area of a two-dimensional surface is directly related to what is going on inside the three-dimensional black hole.

This relationship led to the hypothesis that *the information that describes the interior of a black hole is encoded on the surface of its event horizon.* See Figure 1.20. Some physicists believe that the information defining everything that was sucked into a black hole may be *encoded* on the membrane (surface) of the event horizon. Moreover, they suggest that everything that is sucked into a black hole is returned eventually to the universe, as the black hole gradually evaporates.

As illustrated in Figure 3.7, another hypothesis proposes that information could be stored on the intersecting branes that may be associated with back holes. Figure 3.8 shows how holograms are produced.

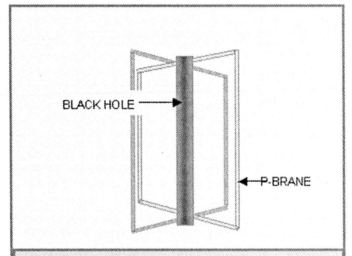

Figure 3.7 Branes and Black Holes. Theorists have hypothesized recently that large p-branes exist in 3-d space. They also speculate that these branes can move through 3-d space and the tiny curled-up shapes. The intersection of two branes may be related to the formation of black holes. It is also believed that information about the contents of black holes may be stored on the surfaces of its intersecting branes.

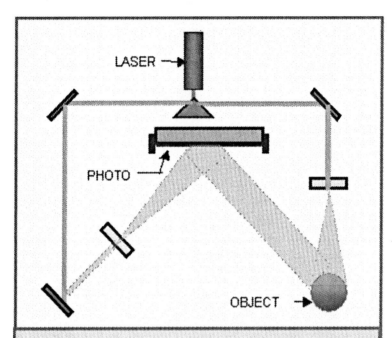

LASER

PHOTO

OBJECT

Figure 3.8 A Holographic Mechanism. One laser beam is split into two beams. The beam on the right passes through a diverging lens and bounces off the object onto a photographic plate. The beam on the left is directed through another diverging lens onto the photographic plate. An interference pattern appears on the developed plate. The 2-d plate records all of the information needed to reproduce a 3-d image of the object. The projection of the image is illustrated below.

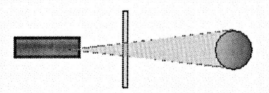

light source   photo plate   3-d object appears in space.

BLACK HOLES AND THE HOLOGRAPHIC PRINCIPLE

While puzzling over the above described relationship and the entropy of black holes, physicists Leonard Susskind and Gerard 't Hooft came up with the *holographic principle*. The principle states that information about the events and objects within a multidimensional volume can be encoded on a lower-dimensional surface. This principle is the basis for holograms. For example, 3-D holographic images are produced from information encoded on the 2-D surfaces of credit cards and transparent photographic plates or film. In principle, the process is rather simple and is illustrated in Figure 3.8.

Holography is related to the interference patterns of light waves. Splitting the light from *one* laser beam into two beams and reflecting them as shown in the figure produces the encoding for holograms. As illustrated at the bottom of Figure 3.8, when a light beam is shined through the developed plate (or film) a complete three-dimensional image of the object(s) appears in space…no screen is needed. A person viewing the projected image can move around it and view all of its sides in detail. (The holographic process can be applied also to motion pictures.)

When done well, these images are amazing. It is more amazing that even a small portion of the photo plate (or film) contains enough information to create the entire three-dimensional image. You probably have realized by now, that the black hole information storage and retrieval systems, described earlier, appear to be analogous to the holographic process.

HOLOGRAPHIC PRINCIPLE AND SS-M THEORY

Well, physicists Suskind and 't Hooft noticed the analogy too, and proposed that the holographic principle applies not only to black holes, but it also applies *to all physical objects and events of our universe.* Soon, theoretical physicists began investigating this intriguing holographic phenomenon. They were amazed to discover that the holographic principle appears to be embedded in SS-M theory.

Currently, this relationship is under intense scrutiny. It appears that SS-M theory has an equivalent theory in which the same physics occurs on a surface with one fewer dimension than the dimensions of the universe that it explains. The holographic principle may be a clue that could lead to the overarching principle currently missing in SS-M theory.

THE EXTENSION OF HUMAN EXISTENCE?

The anthropic principle is often cited as the reason our universe is as we observe and calculate it to be. The principle is stated in different versions, some more forcefully than others. However, it is generally agreed that if our universe didn't have the many properties essential to life, physicists wouldn't be here to try to figure it out. Neither would we. Although it may not seem so at times, overall our universe is "human friendly".

In fact, it may be friendlier than scientists have dared to suspect. With the introduction of strings, branes and the holographic principle we may have a theory broad and powerful enough to support humankind's heretofore *scientifically* impossible dream. That dream seems to be instinctive in humans, regardless of the society or degree of civilization. It is the dream of life after death. Some claim that a mechanism making an afterlife scientifically possible is implicit in the holographic principle. Moreover, the finding that the holographic principle is embedded in the SS-M theory places the holographic principle within the theory.

Humanessence

Let's see if we can logically sketch the framework of a mechanism, within our current scientific knowledge, which theoretically could create an afterlife. As you know, geneticists have mapped the human genome and can relate the genetic code to the living human structure. Thus, there is not only a blueprint for human life, but there is also a mechanism that produces that life.

Although DNA controls the physiology of the whole body, including the human brain, it is not believed that DNA records the *mental essence* of the individual. Probably most of us would agree that the mental essence, as well as the DNA code, determines the definition of a specific individual. Some would argue, however, that the mental essence, alone, contains the *humanessence* of an individual. Since we know how to reproduce the physical individual by cloning, let's focus on humanessence as a critical requirement for the mechanism of an afterlife.

In general, the humanessence of individual humans can only be described on the basis of psychological and physiological observations, including recorded data such as brain wave patterns and imaging. Summarized briefly, the categories of information stored in the brain's memory are: all of the experiences, knowledge, skill patterns, memories and emotions encountered in the lifetime of the individual. In addition, the brain's innate and learned behaviors, including logic and creative thought, the awareness of the existence of self and the ability to relate to exterior sensory input are all a part of humanessence.

We can assume that the brain patterns of humanessence are encoded in certain areas of the brain. Currently, neurological research on the mechanisms of the brain is progressing at an ever-increasing rate. The portions of the brain responsible for handling many of the factors of humanessence have been mapped; and the processing, storage and retrieval of that information are partially understood.

An Afterlife Mechanism

The first step in the afterlife mechanism would be the transfer of the individual's encoded humanessence from the brain to a spatial brane, as indicated by SS-M theory. Since spatial branes are everywhere there is matter, they are present in the matter of the brain. Therefore, it would be expected that an automatic brane encoding process occurs naturally as the information is first encoded electrochemically in the cells of the brain. In other words, as information is electrochemically encoded in the brain, it is encoded also on spatial branes. This step is analogous to light reflecting from an object onto a photographic plate or film, as shown in Figure 3.8.

The second step would take place at the death of the individual. It would consist of the transfer of the humanessence of the individual stored on the branes onto another storage brane or branes, in a place suitable to the continuation of the life of the individual. This process is analogous to the transfer of an image from one photo plate to another.

The last step would be recovery of the humanessence of the individual by a process analogous to projection through the holographic photo plate. That is, the "projection" of energy through the storage brane to reproduce the humanessence of the individual. At this time, we will not consider the question of whether or not a *physical* human being identical to the one that existed on earth is produced. However, it seems logical to assume that an entirely new form, suited to the new environment, *might* be produced.

Well, it seems we have been able to *sketch* a logical, theoretical mechanism that could provide for an afterlife. That is, if all the details fall into place. However, it should be noted that we have assumed nothing about many details, such as the nature and location of the secondary storage branes or the environment of the afterlife.

The secondary storage branes could be located in an extra extended dimension, as may be provided in SS-M theory. In addition, at this time there are a number of theories for other dimensions and other universes that might provide a suitable location for the afterlife.

Although we did not include the DNA of the individual as a part of the humanessence in our mechanism, that question can be left open. If it is nature's way, the DNA could be encoded as a part of the humanessence on the storage brane, without invalidating the general mechanism. It depends on how the process *exists in nature*. At this stage of our hypothesis, we don't have enough information to determine which approach nature would use.

However, no matter how improbable or ridiculous it may seem, if SS-M theory is proven; and if it has the features mentioned earlier; we can expect the afterlife hypothesis to be investigated scientifically. The physics of the afterlife mechanism would certainly be tested vigorously. Eventually, perhaps in this century, there will be an objective scientific consensus on the question. If that consensus should be positive, it probably would be considered to be the most important discovery of all time. It certainly would be the most controversial.

Just how probable is it that such an afterlife mechanism is a part of nature? At this time, no one knows. However, some of us remember when, although they understood the basic science on which television was to be based, a number of scientists didn't believe television would actually work, because it would be too complicated.

Complexity doesn't seem to be a problem these days...just look again at the electronic devices all around you. Remember, however, that these devices were not made possible by the ingenuity of scientists and technologists alone. First, they were made possible by the laws of physics, which are also the *laws of nature*. In addition, the laws of nature *make possible* many things far more amazing than the puny devices of our civilization. For examples just look at some photographs from the HST and recall our discussions as we examined the structure and evolution of our *astonishing* universe.

A BRIEF SUMMARY AND AN UP-DATE

We have followed the evolution of our universe from the big bang to the present...a little more than fourteen billion years. We have seen how a theory based on the simple concept of tiny vibrating strings has the power to explain our universe, and everything in it, better than any other theory. Although not yet complete, SS-M theory can explain much of how and why our universe is as it is.

The recent addition of branes, of p-dimensions, greatly expanded the breadth and potential of the theory.

Recent Findings About Space and Time

Currently many physicists are working diligently to better understand some of the strange properties and concepts emerging from SS-M theory. We have seen a glimpse of one of them...the holographic principle. Others involve the changing concepts of space and time. For example, it has been discovered that space and time behave very strangely in string theories. This really should be no surprise, since strings operate in the bizarre realm of quantum mechanics.

A related discovery is that string theory can be expressed *without space or time*. This leads to the conclusion that even space and time may not be fundamental. A number of physicists are focusing on this question, and it is believed that their research will lead to a better understanding of the *intrinsic* properties of space and time. In any case, SS-M theory still works quite well with our current understanding of space and time.

Everything From Nothing?

Matter and energy are on a better footing. We can manipulate them and even change one into the other. This doesn't mean, however, that they are truly fundamental, since both operate in space and time. The two dual fundamental components (space-time and matter-energy) that we identified as fundamental in the beginning of this book eventually may be relegated to a less fundamental category.

Already, a candidate for that more fundamental component of matter and energy has been proposed...*nothing*. Yes, the nothing of "the void of nothingness" that is hypothesized as the only thing that existed before the nugget and the big bang. You will recall too, that *nothing* is also assumed to be present in the vacuum of empty space.

However, because of quantum fluctuations, *nothing* (zero) may be just the average of *something*. That is, as is the case with virtual particles and energy fluctuations, *something* may just *average* to zero *(nothing)*. In any case, believe it or not, some theoretical physicists are considering the possibility that *nothing* is the one most fundamental component of our universe; and that the "something" (matter and energy), somehow, average to equal *nothing*.

Did Our Universe Arise From Nothing?

In recent years several cosmologists have proposed that the big bang erupted from nothing. In 1973, Edward Tyron suggested that the big bang could have been produced by a random fluctuation in the "vacuum of nothingness". The idea lay dormant for several years.

Then in 1979, Alan Guth came up with his grand inflation theory. You will recall that in this theory a brief very rapid inflation explained the common temperature that now exists throughout our universe. Soon, it was realized that rapid inflation also could explain why omega is one. Furthermore, Guth realized that the rapid expansion would force omega toward one. In fact, calculations indicated that one-second ATB omega would have been *almost* exactly one. Gradually, the above key ideas, considered together, began to paint a more complete picture of the origin of our universe.

The vacuum fluctuation hypothesis soon began to receive a lot of attention from theorists. Recall that we discussed this in our study of quantum mechanics. We used the example of an electron-positron pair suddenly appearing and then annihilating each other almost instantly. Note that in this process, the conservation of matter-energy is maintained.

Many theoretical physicists believe that in addition to the vacuum, a strange *false vacuum* can occur in the quantum realm. Although unfamiliar to us, the properties of a false vacuum seem to be ideal for producing an inflationary universe. For instance, a false vacuum has repulsive gravity. In addition, as it expands, the energy density within increases. Thus, it is argued, that soon after the big bang, as the false vacuum expanded, its energy became dense enough to begin producing the matter particles of our universe. Then the evolution of our universe continued evolving pretty much as indicated in Section 1.

Beginning with a false vacuum and including Guth's grand inflation, we have a theory that appears to explain several recent puzzles of our universe. One is the *increasing* speed of expansion. The false vacuum has built-in repulsive gravity, which may be operating to increase the expansion rate. This repulsive gravity also may be helping to maintain the mass-gravity ratio required to keep omega one. The false vacuum expansion theory also appears to respect the law of conservation of matter-energy. Since matter-energy is positive and the energy of regular gravity is negative, they could cancel each other. That is, if they are actually equal. Recent calculations seem to indicate that they are equal. Therefore, with this theory we can conclude that our universe is made of nothing. However, as we have learned, nothing is actually something that *averages* to nothing.

If it bothers you a bit to discard the nugget as the beginning point of our universe, you are not alone. However, that is the way of science; and it is a major difference between science and dogma. As science progresses, and you continue to learn about theories on the leading edge, you will adjust your understanding of our universe, just as scientists do. You may even accept the concept of a false vacuum someday.

CONCLUSION

Well, I have led you to the edge of the knowledge of our universe; and now I must leave you there. You should consider this little book to be only an introduction to, and overview of, the structure and composition of *our* universe. However, as you just learned, it may all be "much ado about *nothing*". Sorry about that.

Some Information Sources

A FEW LEARNING TIPS

- Examine the *Table of Contents* before beginning the book. The long-form *Table of Contents* provides an outline of the book. The headings for the sections, chapters and topics indicate the interrelationships between the topics and the major concepts to be developed. In addition, the *Table of Contents* is helpful for locating topics for review.

- Before beginning, read the *Preface,* especially the discussion about space particles.

- Use the *Glossary,* as you read, when you don't understand key words. In general, science words are defined when first used either in context or in a statement, but the Glossary definition may help also.

- Don't expect to understand difficult concepts, such as the Heisenberg uncertainty principle, at first reading. Such concepts develop rather slowly, over time. Some concepts are so bizarre that it takes time for the mind to wrap around them.

  Don't be discouraged if, at first, you don't comprehend a principle of quantum mechanics. Most of us didn't understand many of the quantum concepts when first encountered. Moreover, most of us still don't have a full understanding of quantum theory. However, if you persevere, you will understand enough to develop a basic concept of quantum theory and its relationship to our universe.

- As you read, take time to develop an awareness of the extremes of sizes in our universe. The sizes encountered are so extreme that they are just about impossible to imagine. However, you can develop a *relative* under-standing of these sizes.

- If you are not familiar with scientific notation, study the examples given in the *Glossary* under "powers of ten".

INTERNET WEB SITES

Brookhaven National Laboratory. New York. The major facility at the BNL is the Relativistic Heavy Ion Collider (RHIC). A number of experiments can be done with this equipment. In 2002 full power was reached with two beams of gold ions moving in opposite directions in circular vacuum tubes at near the speed of light. When these beams are induced to smash into each other conditions near those of the big bang are created, producing a plasma-like state of matter. Website: www.bnl.gov

Centre European Recherche Nucleaire (CERN). Geneva, Switzerland. A number of particle accelerators, colliders and detectors of several types are in operation or are being constructed. Some are the Large Electron-Positron Collider (LEP), the Large Hadron (proton-proton) Collider and the CERN-Gran Sasso Neutrino Experiment. Website: welcome.cern.ch/welcome/gateway.html

National Aeronautics and Space Administration. NASA operates a number of websites that offer the public information on just about every aspect of space research. There are hundreds of thousands of images and articles that illustrate and explain data and discoveries obtained by NASA's scientists, research vehicles and instruments. The NASA Homepage presents recent news releases, search and links to other websites. You can link to specific sites of interest such as HST photographs, Chandra X-ray images, and more specific topics such as galaxies, quasars, black holes, supernovae and other good stuff. Home Page Website: www.nasa.gov

BOOKS

Cole, K. C. *A Hole in the Universe How Scientists Peered Over the Edge of the Universe and Found Everything.* New York: Harcourt, Inc., 2000. This is a delightful, easy to read book about *nothing*. It deals with the recent ideas about how the universe and everything in it may have originated with an energy fluctuation in the void of nothingness. If you have a preliminary understanding of particle physics and a bit of cosmology, you will enjoy this book.

Greene, Brian. *The Elegant Universe. Superstrings, Hidden Dimensions, and the Quest for the Ultimate Theory.* New York: W W Norton and Company, 1999. This is an account of the development of particle physics from relativity and quantum mechanics to superstring theories and M theory. Some principles of cosmology are interwoven with the physics, especially as they relate to superstring-M theory. *The Elegant Universe* is an elegant, well-written book. Brian Greene is still contributing to the development of superstring-M theory. He is also involved in producing three new Nova programs on the subject. (Watch for them.) This is an excellent basic book for anyone with a desire to understand what may well be the most important theory…ever. *The Elegant Universe* is the best and most comprehensive book currently available on superstring-M theory for the non-physicists reader.

Hawking, Steven. *The Universe in a Nutshell.* New York City, 10036. Bantam Books, 2001. The high quality of the design, paper and illustrations of this book are striking, for a science book. This, plus the enthusiastic writing style and ability of professor Hawking to explain concepts with clear analogies and illustrations, makes an engrossing book. From the staples of particle physics, relativity and quantum mechanics, we are led through some of the most recent and intriguing ideas. These include: black holes and the loss of information, time travel, the future of biological and electronic life, holograms and branes. Superstrings and M theory are discussed and applied from Hawking's unique and creative perspective.

Kane, Gordon. *Unveiling the Ultimate Laws of Nature.* Cambridge, Massachusetts: Helix Books, Perseus Publishing, 2000. *Supersymmetry* focuses on the application of supersymmetry to the standard theory. The experimental search for proof of supersymmetry in accelerator experiments is described. This leads to discussions of the search for the Higgs boson, the hypothetical particle needed to give mass to matter particles in the supersymmetric standard theory. The book concludes with a discussion of current effective theories, including superstring and M theory, and what Kane calls the primary (ultimate) theory. Since Kane is a particle physicist, *Supersymmetry* gives the reader a perspective on physics from the viewpoint of an experimental physicist.

SCIENCE MAGAZINES

Astronomy. Monthly. Kalmbach Publishing Co. Phone: 800-533-6644. Website: www.astronomy.com

Discover. Monthly. Buena Vista Magazines. Phone: 800-829-9132. Website: www.discover.com

Science News. Weekly. Science Service (non profit). Website: www.sciserv.org

Scientific American. Monthly. 415 Madison Avenue, New York, NY 10017. Website: www.sciam.com

Glossary

Absolute Zero. Absolute zero is the coldest anything can be. It is 0 degrees Kelvin, -273.15 degrees Celsius, or -459.67 degrees Fahrenheit.

Accelerator-Colliders. Large devices that use electric and magnetic fields to control and accelerate charged particles to near the speed of light. Controlled streams of the particles (for example, electrons, positrons, protons, antiprotons or gold ions.) are then made to collide with a selected target. The target can be another stream of particles coming from the opposite direction, causing a head-on collision. On the other hand, the target may be a specific kind of matter, such as graphite or gold.

Amplitude. The height of a wave peak or the depth of a wave trough.

Antiparticle. Fundamental particles with electric and force charges that are opposite those of the "ordinary" fundamental particles. For example, the positron carries one positive charge and is the antielectron of the ordinary electron, which has one negative charge. The antiproton carries a negative charge, opposite that of the proton, which is positive.

Antimatter. Antimatter is matter made of antiparticles. For example, an antihydrogen atom is made of one antiproton in the nucleus with one positron (antielectron) orbiting it. Antimatter has the same force of gravity as ordinary matter, but it has opposite electric charge and opposite weak and strong forces. Antihydrogen has been created in the laboratory.

Anthropic Principle. The concept that we observe the universe as we do because if it were different we wouldn't be here to observe it.

ATB. The acronym for "after the bang" meaning the amount of time since the big bang.

Big Bang. The colossal explosion of the nugget that created our evolving universe. Current estimates indicate that the big bang occurred about 14.7 billion years ago.

Black Hole. Enormous astronomical objects with a colossal mass and, therefore, a tremendous gravitational attraction. Nothing within a black hole's event horizon (range) can escape its gravitational attraction. It is believed that large black holes exist at the centers of quasars and galaxies, as well as at other locations throughout the universe.

Boson. A particle produced by a vibrating string of string theory. The boson has a whole number amount of *spin*. Such particles are also known as force particles or messenger particles.

Bosonic String Theory. The original string theory that produced only bosons.

BPS States. The organizational structure of theories with supersymmetry that gives clues to the identity of particles that can be produced in accordance with the theory.

Brane. The membranes allowed by M theory are called branes. The string can be (but usually isn't) called a one-brane, since it has only one dimension. A two-brane is sometimes called just a membrane. A three-brane has three dimensions, etc. Generalized, a p-brane has p spatial dimensions.

Calabi-Yau Shape. A geometric spatial shape in which the six extra dimensions, required by string theory, are curled-up in a manner indicated by the theory. Unlike the three familiar dimensions, Calabi-Yau shapes are so minute that they are not visible.

Closed String. A string in the shape of a loop. The strings that produce the vibrational patterns of the matter and force particles of string theory are loops.

Conifold Transition. The process that changes a Calabi-Yau shape into a different Calabi-Yau shape. In this process the shape can tear and rapidly repair itself without drastic consequences.

Cosmic Microwave Background Radiation. Radiation that was produced by the big bang, still permeates our universe and is all around us. The radiation began as light, but cooled during its some 14 billion years of travel since the big bang; and it now exists as microwave radiation.

Cosmology. The science of the origin, composition, structure, and evolution of our universe. Cosmology and particle physics are rapidly becoming interrelated. They often share experimental data and discoveries.

Cosmological Constant A constant, *lambda*, that was added by Einstein to his general theory of relatively to explain why the objects of the universe were not being drawn together by their mutual gravity. When it was discovered that the universe was expanding, lambda was assumed to be zero. Lambda is now being considered as a possible factor to explain recent observations that the expansion rate of our universe is increasing.

Coupling Constant. The coupling constant is a factor in the formulas of string theories, which indicates the probability that a string will split apart into two strings, or that two strings will rejoin into one. A coupling constant of less than 1 indicates that approximate calculations for that theory will be reasonably accurate.

Curled-up Dimensions. Curled-up dimensions are spatial dimensions that are wrapped-up into ultramicroscopic shapes. Current superstring theories require six, or possibly seven, curled-up dimensions, plus the three familiar spatial dimensions and one time dimension.

Dimensions. In Euclidean geometry there are three directions in a plane perpendicular to each other, which do not intersect. Riemannian geometry allows more than three dimensions and is used in relativity since relativity includes time as a fourth dimension. Superstring-M theory requires a total of eleven dimensions.

Duality. Two or more apparently different string theories may actually be equivalent in their physics. Such theories are described as having the property of duality.

Electromagnetic Force. The electromagnetic force is the union of the magnetic force and the electric force. This union was discovered and explained by Maxwell in terms of electromagnetic fields. The other forces are gravity and the strong and weak nuclear forces.

Electromagnetic Waves. Waves of electromagnetic energy. Light, infrared rays, x-rays, microwaves, television and radio waves are examples of electromagnetic waves. All electromagnetic waves travel at the same speed in a vacuum: 186,000 miles per second.

Electron. The electron is one of the fundamental particles of matter. The electron carries one negative charge and has a mass of .00054.

Eleven-dimensional Supergravity. One of the early higher-dimensional string theories that may play a role in superstring-M theory.

Entropy. Entropy is the measure of the disorder within a physical system. It can be thought of as the measure of the *number of arrangements* of the smaller objects of a system that can be made without changing the system's larger appearance. Entropy may be considered also to be the information defining the contents of a black hole.

Equivalence Principle. A fundamental principle of general relativity that states the equivalence of accelerated motion and the effects of a gravitational field. This allows all observers to assume they are at rest, as long as they consider that they are in a suitable gravitational field.

Event Horizon. The surface area at the mouth of a black hole is called the event horizon. Once anything, including light, falls through the event horizon it cannot escape the black hole because of the extreme gravity within. (A recent theory indicates that radiation and particles may leak, *very* slowly, out of black holes.)

Extended Dimensions. The three large dimensions of our visible universe. Also, any other unfurled dimensions proposed in some speculative theories of the universe.

Families of Matter Particles. The twelve fundamental particles of matter can be arranged into three groups, or families, of four. The masses of each corresponding particle of each family are different. See Table 1.

Fermions. A fundamental particle, such a matter particle, which has half a whole *odd number* amount of spin.

First Superstring Revolution. The period from 1984 to 1986 when there was a flurry of activity, by physicists throughout the world, to further develop string theory. Their combined efforts demonstrated that many features of the standard theory, including quantum theory, emerge *naturally* from string theory.

Flop Transition. A change in a Calabi-Yau shape into a *different* Calabi-Yau shape.

Force Charge. The property of a particle that causes it to react in a certain way to a specific force. Force charges include the familiar positive and negative electric charges and the less familiar strong and weak charges that operate only at nuclear distances.

Frequency. The number of vibrations (cycles) per second in a waveform.

Gluon. The force particle of the strong force. See Table2.

Gravitational Force. One of the four fundamental forces. On a particle-to-particle basis, gravity is the weakest of the four forces.

Graviton. The graviton is the hypothetical force particle of the gravitational force. The graviton is produced, theoretically, by one of the patterns of vibrating strings and has zero mass. The graviton has not been found experimentally. See Table 2.

Holographic Principle. The concept that the information in a multidimensional system can be stored on a boundary of that system. The boundary will have one less dimension than the system.

Humanessence. The combination of all of the mental ingredients that make up the essence of a human being. The essence of an individual human encoded in the brain.

Infinities. The results of mathematical calculations involving point particles, relativity and quantum mechanics that appear to be nonsense.

Inflation. Our universe is constantly inflating (expanding) as a result of the big bang. It is also believed that just after the big bang there was a brief enormous inflation. This modification to the standard theory of cosmology is known as the inflationary cosmological theory.

Interference Pattern. A pattern formed by the overlapping of waves emitted from different directions or different times.

Kaluza Klein Theory. An Early theory that added extra tiny dimensions to quantum theory. This aided the early development of string theories.

Kelvin Temperature Scale. A temperature scale that starts at absolute zero. Zero degrees K is absolute zero and is equal to—273.15 degrees Celsius or—458.67 degrees Fahrenheit.

Massless Black Holes. The tiny black holes of string theory that become lighter in mass as a portion of Calabi-Yau space shrinks. When they reach point size the black holes no longer exist and are replaced by a massless string.

Maxwell's Electromagnetic Theory. A Theory that unites electricity and magnetism. The theory is based on the concept of the electromagnetic field.

Messenger Particle. The smallest bundle of a force that transmits that force. For example, the photon is the messenger particle for the electromagnetic force.

M Theory. One of the most promising recent theories of particle physics. M theory allows membranes of a number of dimensions and appears to require a total of eleven dimensions, including time. The string, of string theories, is a one-brane under the definition of membranes in M theory.

Neutron. A neutral particle with zero electric charge. Neutrons are usually found with protons in the nuclei of atoms. The mass of the neutron is slightly greater than the mass of the proton.

Neutrino. A neutral particle of almost, but not quite, zero mass. Neutrinos are emitted in colossal numbers from the fusion reactions of stars, including our sun. Since they have a tiny mass, only the weak charge and travel at near the speed of light, they pass straight through almost anything. Huge quantities of neutrinos were produced in the aftermath of the big bang.

Nucleus. The nucleus is the core of an atom. It contains protons, and usually neutrons, which are held together by gluons. The nucleus of the simplest atom, hydrogen, consists of only one proton.

Open Strings. The strings of string theories that have two unconnected ends.

Photoelectric Effect. The ejection of electrons from a metal surface when light strikes the surface. The speed of ejected electrons is determined by the frequency of the light. Higher frequencies eject faster electrons.

Photon. The force particle of the electromagnetic force. The photon may be thought of as the smallest bundle of energy of light and other electromagnetic radiation.

Planck's Constant. The basic constant of quantum mechanics. Planck's constant determines the size of the discrete quanta (units) of mass, energy, spin, etc. of the ultramicroscopic realm of quantum mechanics. It is 1.05×10^{-27} grams-centimeter/second.

Planck Energy. The typical energy of the vibration patterns of strings in string theories. The Planck energy is estimated to be about a thousand kilowatt hours.

Planck Length. About 10^{-33} centimeters. This is the length of a typical string in string theories. It is below this size that the fluctuations in the space-time vacuum become extremely large.

Planck Mass. The mass equivalent of a typical vibrational pattern of a string in string theories. The Planck mass is about ten billion billion times the mass of the proton. That's about the mass of a small dust particle. That's a lot when you consider how unbelievably tiny a string (or the particle it produces) is.

Planck Tension. The tension of the strings in string theories. It is about 10 to the 39th tons.   That's. 10,000,000,000,000,000,000,000,000,000,000,000,00 0. *Tons!*

Planck Time. Planck time is the time when the diameter of our universe was the Planck length. It is about 10^{-43} seconds after the big bang. It is also the time required for light to travel the Planck length.

Powers of Ten. Also called scientific notation. A method of representing large or small numbers with exponents. For example, $10^0=1$, $10^1=10$, $10^2=100$, $10^3=1000$, Using this technique, we can represent very large numbers in a brief manner. Examples are: 10^{12}=*one* trillion. For another example, look at *Planck tension*, above, it can be written as 10^{39} tons. We can use the same method to represent small numbers that are less than one (decimals). This time we use *negative* exponents to the base 10. Examples:10^{-1}=.1, 10^{-2}=.01, 10^{-3}=.001, 10^{-4}=.0001, etc. A smaller example: 10^{-16}=.0000000000000001. That's one *quadrillionth* of something.

Primordial Nucleosynthesis. The formation of nuclei of future atoms by the union of quarks to produce neutrons and protons. This happened during a three-minute period soon after the big bang.

Proton. A particle found in the nuclei of atoms. It is made of two up-quarks and one down-quark; and it has one positive electric charge.

Quark. A fundamental particle that reacts to the strong force. Quarks exist in six varieties with a wide range of masses. Quarks, held together by gluons, make protons and neutrons.

Quanta. The smallest units into which things can be divided according to quantum theory. For example, photons are the smallest packets (quanta) of the electromagnetic field.

Quantum Theory. Laws of physics that are essential at the ultramicroscopic level of the fundamental particles, their forces and energy charges. These laws deal with such concepts as wave-particle duality, Heisenberg uncertainty and quantum fluctuations.

Quantum Tunneling. A prediction of quantum theory that there is a small probability that objects can pass through barriers that should stop them, according to common experience and the classical laws of physics.

Radiation. Energy, such as light or radio waves, carried by electromagnetic waves. Radioactivity can also include fast moving particles such as electrons (beta rays) as well as short electromagnetic waves (gamma rays).

Resonance. A natural state of vibration of a vibrating object, such as a string.

Reciprocal. The inverse of a number. For example, the reciprocal of 4 is ¼, and the reciprocal of 1/4 is 4.

Relativity. There are two relativity theories. *Special relativity* declares that all observers in a state of *constant velocity* are subject to the same laws of physics. This means that all constant-velocity observers can claim that they are at rest. In addition, all observers can claim to be at rest regardless of their state of motion if they recognize that they are in a gravitational field. *General relativity* states that the gravitational force is communicated through the curvature of spacetime.

Second Superstring Revolution. A period, beginning in 1995, during which many physicists were encouraged by the realization that there might be a way to proceed in string theory without the needed unknown equations. They recognized that duality, and other properties of string theories, would enable them to determine certain properties without the equations. These methods continue to be helpful.

Singularity. A point in spacetime where a catastrophic rip occurs or is predicted to occur. Superstring theory indicates a way that such tears may be repaired as they are happening, thus avoiding a catastrophe.

Standard Theory of Particle Physics. The standard theory (or model) evolved along with the discoveries in quantum mechanics during the last century. Standard theory treats the particles of matter as point particles. It does not include the gravitational force or general relativity. The theory has been very successful in explaining experimental data arising from quantum mechanics. However, physicists must turn to general relativity to explain our larger universe.

String. A one-dimensional "sliver of space" that is the fundamental unit of string theories.

String Mode. A configuration that a string may assume. For example, a string may assume a vibrational mode or winding mode.

String Theory. Any of a number of theories based on tiny strings as the fundamental particles of matter and the forces. String theory unifies general relativity and quantum mechanics. String theory has been evolving since about 1968, and the later versions are known as superstring theories. Superstring theory is a unified theory that deals with all of the physics of nature.

Strong Force. The force that binds quarks inside protons and neutrons. The strong force also holds protons and neutrons inside the nuclei of atoms. The strong force particle is the gluon.

Superparticles. See superpartners, below.

Superpartners. Partners of the fundamental particles of matter and the forces. They differ from the familiar particles by one-half unit of spin. Superpartners are required by supersymmetry. They are also referred to as superparticles.

Superstring Theory. Any string theory that includes supersymmetry in its structure.

Superstring-M Theory. A title used to represent the theory that is evolving from a combination of elements of superstring and M theories.

Supersymmetric Standard Theory. The standard theory modified by the addition of supersymmetry.

Supersymmetry. A type of symmetry that interrelates the properties of particles with a whole even number amount of spin (bosons) to those with half a whole *odd* number amount of spin (fermions).

Symmetry. A spacetime property of a physical system that stays the same when the system is changed in some way or the system is observed from a different perspective (angle). For example, a sphere has rotational symmetry since it will appear the same when rotated around any axis or viewed from any angle.

Symmetry Breaking. A reduction in the amount of symmetry in a system. This is usually brought about by a change of phase. An example is the phase change of water to ice. Symmetry breaking occurs frequently in the evolution of our universe.

Symmetry Building. The opposite of symmetry breaking. An example of symmetry building is the melting of ice. Note that in symmetry breaking, the energy of the systems is reduced; and in symmetry building the energy of the system is increased.

Torus. The surface of a donut shape.

Ultramicroscopic. As used in this book *ultramicroscopic* refers to sizes a little above and below the Planck length.

Uncertainty Principle. A major principle of quantum theory. The principle states that there are measurements that cannot be known with complete precision. For example, the position and velocity of a particle, at a specific moment, cannot be known with precision. This uncertainty and its consequences become greater as the event is considered at smaller and smaller distance and time scales.

Unified Theory. A theory that unifies all four forces and the particles of matter in one theoretical structure. Superstring theory appears to accomplish this unification.

Vibration Pattern. The amplitude, and number of peaks and troughs per second of a vibrating string.

Virtual Particles. Particles that appear in the vacuum for an instant and then quickly annihilate. Virtual particles are formed from energy borrowed from the universe and repaid *almost* instantly.

W boson. One of two force particles of the weak force. The other is the Z boson.

Wave Function. The probability waves of quantum theory.

Wave Length. The distance from one peak or trough of a wave to the next peak or trough.

Wave-particle Duality. The principle that states that electromagnetic waves have particles associated with them. The inverse also appears to be true: particles have waves associated with them.

Weak Force. One of the four fundamental forces. The weak force particles are the W and Z bosons. The weak force is associated with the decay of radioactive elements and the interaction of neutrinos with other particles.

Winding Energy. The energy of a string wound around a circular spatial dimension.

Winding Mode. The pattern of a string that wraps around a circular spatial dimension.

Winding Number. Indicates the number of times a string is wound around a circular spatial dimension.

Z boson. One of two force particles of the weak force. The other is the W boson.

0-595-27589-3

Printed in the United States
15658LVS00008BA/274-279